HVDC and FACTS Controllers

Applications of Static Converters in Power Systems

POWER ELECTRONICS AND POWER SYSTEMS

Series Editors
M. A. Pai and Alex Stankovic

Other books in the series:

POWER QUALITY ENHANCEMENT USING CUSTOM POWER DEVICES
Arindam Ghosh, Gerard Ledwich, 1-4020-7180-9
COMPUTATIONAL METHODS FOR LARGE SPARSE POWER SYSTEMS ANALYSIS: *An Object Oriented Approach,*
S. A. Soman, S. A. Khaparde, and Shubha Pandit, ISBN 0-7923-7591-2
OPERATION OF RESTRUCTURED POWER SYSTEMS
Kankar Bhattacharya, Math H.J. Bollen and Jaap E. Daalder, ISBN 0-7923-7397-9
TRANSIENT STABILITY OF POWER SYSTEMS: *A Unified Approach to Assessment and Control*
Mania Pavella, Damien Ernst and Daniel Ruiz-Vega, ISBN 0-7923-7963-2
MAINTENANCE SCHEDULING IN RESTRUCTURED POWER SYSTEMS
M. Shahidehpour and M. Marwali, ISBN: 0-7923-7872-5
POWER SYSTEM OSCILLATIONS
Graham Rogers, ISBN: 0-7923-7712-5
STATE ESTIMATION IN ELECTRIC POWER SYSTEMS: *A Generalized Approach*
A. Monticelli, ISBN: 0-7923-8519-5
COMPUTATIONAL AUCTION MECHANISMS FOR RESTRUCTURED POWER INDUSTRY OPERATIONS
Gerald B. Sheblé, ISBN: 0-7923-8475-X
ANALYSIS OF SUBSYNCHRONOUS RESONANCE IN POWER SYSTEMS
K.R. Padiyar, ISBN: 0-7923-8319-2
POWER SYSTEMS RESTRUCTURING: *Engineering and Economics*
Marija Ilic, Francisco Galiana, and Lester Fink, ISBN: 0-7923-8163-7
CRYOGENIC OPERATION OF SILICON POWER DEVICES
Ranbir Singh and B. Jayant Baliga, ISBN: 0-7923-8157-2
VOLTAGE STABILITY OF ELECTRIC POWER SYSTEMS
Thierry Van Cutsem and Costas Vournas, ISBN: 0-7923-8139-4
AUTOMATIC LEARNING TECHNIQUES IN POWER SYSTEMS,
Louis A. Wehenkel, ISBN: 0-7923-8068-1
ENERGY FUNCTION ANALYSIS FOR POWER SYSTEM STABILITY
M. A. Pai, ISBN: 0-7923-9035-0
ELECTROMAGNETIC MODELLING OF POWER ELECTRONIC CONVERTERS
J. A. Ferreira, ISBN: 0-7923-9034-2
SPOT PRICING OF ELECTRICITY
F. C. Schweppe, M. C. Caramanis, R. D.Tabors, R. E. Bohn, ISBN: 0-89838-260-2
THE FIELD ORIENTATION PRINCIPLE IN CONTROL OF INDUCTION MOTORS
Andrzej M. Trzynadlowski, ISBN: 0-7923-9420-8
FINITE ELEMENT ANALYSIS OF ELECTRICAL MACHINES
S. J. Salon, ISBN: 0-7923-9594-8

HVDC and FACTS Controllers

Applications of Static Converters in Power Systems

by

Vijay K. Sood, Ph.D., ing., FEIC

KLUWER ACADEMIC PUBLISHERS
Boston / Dordrecht / New York / London

Distributors for North, Central and South America:
Kluwer Academic Publishers
101 Philip Drive
Assinippi Park
Norwell, Massachusetts 02061 USA
Telephone (781) 871-6600
Fax (781) 871-6528
E-Mail: <kluwer@wkap.com>

Distributors for all other countries:
Kluwer Academic Publishers Group
Post Office Box 322
3300 AH Dordrecht, THE NETHERLANDS
Telephone 31 78 6576 000
Fax 31 78 6576 254
E-Mail: <services@wkap.nl>

Electronic Services <http://www.wkap.nl>

Library of Congress Cataloging-in-Publication Data

HVDC and FACTS Controllers: *Applications of Static Converters in Power Systems*
Vijay K. Sood
ISBN 1-4020-7890-0
eBook ISBN: 1-4020-7891-9

Printed on acid-free paper.

Printed in the United States of America

This book is dedicated to my family for their ever loving support over the years. My thanks to my wife Vinay, daughter Nishi and son Ajay.

Contents

Preface

The motivation to write this book came from the observation that the technology of High Voltage DC (HVDC) Transmission and Flexible AC Transmission Systems (FACTS) has evolved much over the past 10 years, and no recent book addressed the developments in these areas in a compact, easily available and readable manner. These new developments came about primarily due to the advances in power electronic switches and microprocessors. This new book hopes to address a small part of the void that existed in the technical book area.

Some words should be said about the title of the book: *HVDC and FACTS controllers*. Modern thinking and terminology usage dictates that HVDC transmission should now be considered as a part of FACTS controllers. However, a look at the evolution in the field shows that the technology started with HVDC transmission and FACTS controllers came along much later. Hence my choice of the title reflects this.

The pioneering books in the field of HVDC and FACTS by well-known authors such as E.W.Kimbark, J.Arrillaga, K.R.Padiyar, N.G.Hingorani and L.Gyugyi were a source of great inspiration to me and will unmistakably have been reflected in the work here ... in some shape or form. I have also referred to many documents and papers from the archives of the IEEE, and this preface would be incomplete without acknowledging these reference sources.

The book writing process often has its own agenda. Furthermore, it is a difficult and time consuming affair. In this case the original, planned for time duration had to slip by many months due to personal health problems and other time commitments. Consequently, the original outline for the content and extent had to be altered somewhat from the final submission. However, I do thank the support and patience of Alex Greene and Melissa Sullivan at Kluwer Academic Publishers for sticking with me during this difficult period. I thank also Dr. M.A.Pai who provided technical support and guidance during the feasibility and review process of the book.

The book is intended for senior under-graduate students, graduate students and practising utility engineers. Consequently, mathematical/theoretical explanations have been kept to a minimum, and a focus on practical issues was maintained throughout the book.

This book would not have been possible without the efforts and contributions of many individuals. These individuals can be split into two groups: my former and present students at Concordia University, Montreal and many of my colleagues at IREQ (Hydro-Québec). These individuals are true troopers and I thank them for their support, camaraderie and skills over the many years that I had the pleasure of working with them.

My affiliation with the electrical engineering department at Concordia University dates back to 1984 when I started as an Adjunct Assistant Professor. The work of some of my former students figures prominently in this text book and I would like to thank Vijay Khatri, Krish Narendra, Rachit Arora and Anup Mazumder for their contributions.

I have spent most of my working life of 28 years plus, at the research institute of Hydro-Québec (known internationally as IREQ), mostly in the Power Systems Simulation Group, but more recently in the Electrical Equipment Group. I would like to thank my former colleagues P.C.S. Krishnayya, Harbans Nakra, Lewis Vaughan, Ly Bui and David McCallum amongst others. It is not possible to name individually all the other colleagues at IREQ with whom I have inter-acted with over the years, but I do thank them collectively. In addition, there are many other individuals at Hydro-Québec, and elsewhere, who have contributed, in their own way, to the subject matter at hand.

Other associates who have provided indirect moral support and inspiration over the years are Professors Loi Lei Lai at City University, London (UK) and H.S.Chandrasekhariah of IISc, Bangalore (now deceased).

I had considerable support from manufacturers and other corporate sponsors in the preparation of this book. They assisted primarily with the supply of both technical literature and photographs. These individuals include: Niclas Ottosson and Lewis Vaughan (ABB), Peter Lips (Siemens), Michael Baker (Alstom), Paul Wilson (Manitoba HVDC Research Centre), members of RTDS Technologies Inc. and Bahram Khodabakhchian (TransÉnergie Technologies). My grateful appreciation to all of them.

The book starts with a short Introduction to HVDC transmission. This is followed by chapter 2 with a discussion of the two types of converters - namely current and voltage source converters - which form the core converters for applications in HVDC and FACTs controllers. The current source converter has traditionally been associated with HVDC transmission. However, the more recent development of the voltage source converter has extended the reach and opened new areas of applications. Chapter 3 deals with synchronization techniques for the power converters with the ac system frequency. This topic has not been given the necessary importance in previous textbooks and is likely to gain in importance in the future as (a) inter-connections to weak ac systems become more frequent, and (b) inter-actions between ac-dc-ac systems become more relevant. Chapter 4 deals with HVDC control strategies for two-terminal HVDC systems; an extension to multi-terminal HVDC systems had to be cut due to time limitations. Chapter 5 deals with a fundamental look at forced commutated converters, and lays out some of the basic considerations. With the availability of new switching devices, some of these considerations will take on added importance in the future. The recent success of the capacitor commutated converter is highlighted in chapter 6. This new application of an old configuration demonstrates the need for persistence and research effort needed to develop ideas to fruition. Chapter 7 describes a new converter configuration and a FACTS application which promises an exciting future for this technology. Some of the persistent research efforts have led to new applications and possibilities for using the voltage source converter in future HVDC applications and these are highlighted in chapter 8. The new control paradigms developed in the past ten years have opened the door to exciting new applications of active ac and dc filters. Chapter 9 presents one of these applications and shows the control paradigms required. Much research in this area is on-going and the future holds new applications and reduced costs for the converters. Chapter 10 presents the results of an EMTP based study of typical disturbances on an HVDC system. This demonstrates the need for optimization of controller gains and effort and detail required to do studies to obtain the gain parameters of the controllers. The use of new control algorithms using fuzzy logic, neural networks and intelligent controllers may ease some of the burden placed on existing PI controllers; this is a topic of intense research and will figure prominently in the future. Chapter 11 deals with a simple neuro-fuzzy adaptive controller which could be used to enhance performance of a HVDC system. Chapter 12 is a short introduction to measurement and monitoring aspects related to the controllers. Chapter 13 deals with some of the studies that the author has been associated with and brings out the inter-action aspects of ac-dc systems and controllers. Considerable number of studies of this nature have to be carried out by the practising engineer to verify system behavior and stability. No mathematical development into the topic is presented and will be left to another project in the future. Some of the tools that are used by practising engineers in the afore-mentioned studies are presented in Chapter 14. In the final chapter, a look at the modern HVDC system where the focus will be on cost reduction using the latest techniques is presented.

Finally, I would like to emphasize the solid faith, effort and contributions of my wife Vinay in the preparation of this book. She was a constant source of support and provided many valuable contributions in typing and re-typing the manuscript, drawing the figures, proof reading and doing the necessary corrections. Her patience in tolerating the many hours that I was away tackling this work while neglecting my other duties is most admirable.

Vijay K. Sood
Brossard, Québec.
January, 2004
v.sood@ieee.org

Acronyms

ACF	AC Filter
A/D Converter	Analog to Digital Converter
AF	Active Filters
ALPRET	Alpha Retard (at rectifier)
ATP	Alternative Transients Program (a freeware version of EMTP)
BB	Back to Back (Converter)
CC	Circuit Commutation
CC	Current Control
CCC	Capacitor Commutated Converter
CCO	Current Controlled Oscillator
CE	Current Extinction
CEA	Constant Extinction Angle
CF	Commutation Failure
CLC	Chain Link Converter
CSC	Current Source Converter
CSCC	Controlled Series Capacitor Converter
ConTune	Continuously Tuned (Filter)
DCF	DC Filter
DCG	Development Co-Ordination Group (of EMTP)
DCPT	DC Potential Transformer
DSP	Digital Signal Processor
DQO	Direct Quadrature Zero
EHV	Extra High Voltage
ESCR	Effective Short Circuit Ratio
EMTP	Electro-Magnetics Transients Program
EMTPWorks RV	New version of EMTP Restructured Version
EMTDC	Electro-Magnetics Transients for DC
EPC	Equi-Distant Pulse Control
EMI	Electro-Magnetic Interference
FC	Forced Commutation
FACTS	Flexible Alternating Current Transmission Systems
FR	Force-Retard

GFU	Grid Firing Unit
GTO	Gate Turn-Off (Thyristor)
GUI	Graphical User Interface
HB	Hysteresis Band
HVDC	High Voltage Direct Current
HVDC Light	Trade mark of a version of HVDC converter for low powers
HYPERSIM	Trade mark of a version of digital simulator
IGBT	Insulated Gate-Bipolar Transistor
IPC	Individual Phase Control
IPP	Inter-Mountain Power Project
IOLIM	Current Order Limited
IMIN	Current Order Minimum
LCC	Line Commutated Converter
LC	Line Commutation
LPOF	Low Pressure Oil Filled
LTT	Light-Triggered Thyristor
LU	Lower Upper
MAP	Maximum Available Power
MTDC	Multi-terminal Direct Current
MVA	Mega Volt Amps
MCT	MOS Controlled Transistor
MOSFET	MOS Field Effect Transistor
MP	Mid Point
MISO	Multiple-Input Single-Output
MODELS	New EMTP subroutine for controls
OCP	Over Current Protection
OOM	Object Oriented Method
OCT	Optical Current Transducer
PLL	Phase Locked Loop
PWM	Pulse Width Modulation
PFC	Pulse Frequency Control
PPC	Pulse Phase Control
PS	Periodical Sampling
PPCO	Pulse Phase Control Oscillator
PDP	Pole Differential Protection
PT	Potential Transformer
PSCAD	Power System Computer Aided Design

RoW	Right of Way
RBF	Radial Basis Function
RBFNN	RBF Neural Network
RLC	Resistor, Inductor and Capacitor
RV	Restructured Version
SIL	Surge Impedance Loading
SI	Static Induction Thyristor
SVC	Static Var Compensators
SCR	Short Circuit Ratio
STATCOM	Static Compensator
SLG	Single Line to Ground (Fault)
THD	Total Harmonic Voltage Distortion
TNA	Transient Network Analyzer
TC	Triangular Carrier
UHV	Ultra High Voltage
VC	Voltage Control
VH	Valve Hall
VY	Valve Yard
VAR	Volt Amps reactive
VCO	Voltage Controlled Oscillators
VDCL	Voltage Dependent Current Limit
VDCOL	Voltage Dependent Current Order limit
VSC	Voltage Source Converter
VGP	Valve Group Protection
XLPE	Cross Linked Poly-Ethylene
$HVDC^{PLUS}$	HVDC Power Link Universal System from Siemens
NF	Neuro-Fuzzy

Corporate entities

CEPEL	Name of research organization (of Brazil)
CESI	Name of research organization (of Italy)
CPRI	Central Power Research Institute (of India)
ABB	Manufacturer of electrical equipment
Siemens	Manufacturer of electrical equipment
CIGRÉ	*Conseil International des Grands Réseaux Électriques*
IEEE	Institute of Electrical and Electronics Engineers
BPA	Bonneville Power Administration
IREQ	Institute of research of Hydro-Québec, Varennes, Québec.
RTDS	Real Time Digital Simulator Technologies Inc.

Chapter 1

Introduction to HVDC Transmission

1.1 INTRODUCTION

Although earlier books on HVDC Transmission [1,2,3] have already covered adequately an introduction to HVDC technology, a brief introduction here is deemed necessary for the sake of completeness. This presentation will likely be considered a repeat of old material since the technology is now considered mature. Therefore, this chapter will be brief as the book's main focus is on control aspects.

HVDC technology first made its mark in the early under-sea cable interconnections of Gotland (1954) and Sardinia (1967), and then in long distance transmission with the Pacific Intertie (1970) and Nelson River (1973) schemes using mercury-arc valves. A significant milestone occurred in 1972 with the first Back to Back (BB) asynchronous interconnection at Eel River between Quebec and New Brunswick; this installation also marked the introduction of thyristor valves to the technology and replaced the earlier mercury-arc valves.

The first 25 years of HVDC transmission were sustained by converters having mercury arc valves till the mid-1970s. The next 25 years till the year 2000 were sustained by line-commutated converters using thyristor valves. It is predicted that the next 25 years will be dominated by force-commutated converters [4]. Initially, this new force-commutated era has commenced with Capacitor Commutated Converters (CCC) eventually to be replaced by self-commutated converters due to the economic availability of high power switching devices with their superior characteristics.

The growth of HVDC transmission is clearly visible in a listing of HVDC projects prepared by the IEEE Transmission and Distribution Committee (Table 1-1) [5].

Table 1-1: Listing of HVDC Installations

Link	HVDC Supplier	Year	Rating (MW)	DC Voltage (kV)	Line &/or Cable (km)	Location
Gotland I#	A	1954	20	+/- 100	96	Sweden
English Channel	A	1961	160	+/- 100	64	England-France
Volgograd-Donbass*		1965	720	+/- 400	470	Russia
Inter-Island	A	1965	600	+/-250	609	New Zealand
Konti-Skan I	A	1965	250	250	180	Denmark-Sweden
Sakuma	A	1965	300	2x125	B-B***	Japan
Sardinia	I	1967	200	200	413	Italy
Vancouver I	A	1968	312	260	69	Canada
Pacific Intertie	JV	1970	1440	+/-400	1362	U.S.A.
		1982	1600			
Nelson River I**	I	1972	1620	+/-450	892	Canada
Kingsnorth		1975	640	+/-260	82	England
	I					
Gotland	A	1970	30	+/-150	96	Sweden
Eel River	C	1972	320	2x80	B-B	Canada
Skagerrak I	A	1976	250	250	240	Norway-
II	A	1977	500	+/-250		Denmark
III	A	1993	440	350	240	Norway-Denmark
Vancouver II	C	1977	370	-280	77	Canada
Shin-Shi-nano	D	1977	300	2x125		Japan
		1992 est.	600	3x125	B-B	
Square Butte	C	1977	500	+/-250	749	U.S.A.
David A. Hamil	C	1977	100	50	B-B	U.S.A.
Cahora Bassa	J	1978	1920	+/-533	1360	Mozambique-S.Africa
Nelson River II	J	1978	900	+/-250	930	Canada
		1985	1800	+/-500		
CU	A	1979	1000	+/-400	710	U.S.A.
Hokkaido-Honshu	E	1979	150	125	168	Japan
	E	1980	300	250		

Table 1-1 (cont.):Listing of HVDC Installations

Link	HVDC Supplier	Year	Rating (MW)	DC Voltage (kV)	Line &/or Cable (km)	Location
		1993 est.	600	+/-250		
Acaray	G	1981	50	25.6	B-B	Paraguay
Vyborg	F	1981	355	1x170(+/-85)	B-B	Russia-Finland
	F	1982	710	2x170		
			1065	3x170		
Duernrohr	J	1983	550	145	B-B	Austria
Gotland II	A	1983	130	150	100	Sweden
Gotland III	A	1987	260	+/-150	103	Sweden
Eddy County	C	1983	200	82	B-B	U.S.A.
Château-guay	J	1984	1000	2x140	B-B	Canada
Oklaunion	C	1984	200	82	B-B	U.S.A.
Itaipu	A	1984	1575	+/-300	785	Brazil
	A	1985	2383			
	A	1986	3150	+/-600		
Inga-Shaba	A	1982	560	+/-500	1700	Zaire
Pacific Intertie up-grade	A	1984	2000	+/-500	1362	U.S.A.
Blackwater	B	1985	200	57	B-B	U.S.A.
Highgate	A	1985	200	+/-56	B-B	U.S.A.
Madawaska	C	1985	350	140	B-B	Canada
Miles City	C	1985	200	+/-82	B-B	U.S.A.
Broken Hill	A	1986	40	2x17(+/-8.33)	B-B	Australia
Intermoun-tain Power Project	A	1986	1920	+/-500	784	U.S.A.
Cross-Channel (Les Man-darins) (Sellindge)						
	H	1986	1000	+/-270	72	France-
	I	1986	2000	2x+/-270		England
Des Can-tons-Com-erford	C	1986	690	+/-450	172	Canada-USA.
Sacoi ##	H	1986	200	200	415	Corsica Island
###		1992 est.	300			Italy
Itaipu II	A	1987	3150	+/-600	805	Brazil
Sidney (Virginia Smith)	G	1988	200	55.5	B-B	U.S.A.
Gezhouba-Shanghai	B+G	1989	600	500	1000	China
		1990	1200	+/-500		
Konti-Skan II	A	1988	300	285	150	Sweden-Denmark

Table 1-1 (cont.):Listing of HVDC Installations

Link	HVDC Supplier	Year	Rating (MW)	DC Voltage (kV)	Line &/or Cable (km)	Location
Vindhy-achal	A	1989	500	2x69.7	B-B	India
Pacific Intertie expansion	B	1989	1100	+/-500	1362	U.S.A.
McNeill	I	1989	150	42	B-B	Canada
Fenno-Skan	A	1989	500	400	200	Finland-Sweden
Sileru-Bar-soor	K	1989	100	+100	196	India
			200	+200		
			400	+/-200		
Rihand-Delhi	A	1991	750	+500	910	India
		1991	1500	+/-500		
Hydro Que-bec-New England	A	1990	2000****	+/-450	1500	Canada-USA.
Nicolet tap	A	1992	3000			Canada
Welch-Monticello		1995 est.	600		B-B	USA.
Etzenricht	G	1993 est.	600	160	B-B	Germany-Czech
Vienna South-East	G	1993 est.	550	145	B-B	Austria-Hungary
DC Hybrid link	AB	1992 est.	992	+270/-350	617	New Zealand
Chan-drapur-Padghe		1997 est.	1500	+/-500	900	India
Chan-drapur-Ramagun-dum		1996 est.	1000	2x	B-B	India
Gazuwaka-Jeypore		1997 est.	500		B-B	India
Leyte-Luzun		1997 est.	1600	400	440	Philippines
Haenam-Cheju	I	1993 est.	300	+/-180	100	South Korea
Baltic Cable Project	AB	1994 est.	600	450		Sweden-Germany
Victoria-Tasmania			300	30		Australia
Kontek HVDC Intercon.		1995 est.	600	400		Denmark
Scotland-N. Ireland		1996 est.	250	250		United Kingdom
Greece-Italy		1997 est.	500			Italy

Table 1-1 (cont.):Listing of HVDC Installations

Link	HVDC Supplier	Year	Rating (MW)	DC Voltage (kV)	Line &/or Cable (km)	Location
Tsq-Beijao		1997 est.	1800	500	903	China
Iceland-Scotland link			2000		950	
Sarawak-Malaysia			1600		620	

Notes for Table 1-1.

A - ASEA
B - Brown Boveri
C - General Electric
D - Toshiba
E - Hitachi
F - Russian
G - Siemens
H - CGEE Alsthom
I - GEC (Formerly English Electric)
J - HVDC W.G. (AEG, BBC, Siemens)
K - (Independent)
AB - ABB Brown Boveri
JV - Joint Venture (GE and ASEA)

*	2 V.G.'s replaced with thyristors in 1977
**	2 V.G.'s in Pole 1 replaced with thyristors by GEC in 1991.
***	Back-to-back HVDC System
****	Multi-terminal system. Largest terminal is rated 2250 MW.
##	Retired from Service
##	50 MW thyristor tap
####	Up rated w/thyristor valves

To understand this impressive growth of HVDC transmission in the past 50 years, it is first necessary to compare it to conventional ac transmission.

1.2 COMPARISON OF AC-DC TRANSMISSION

An evaluation of transmission costs, technical considerations, and the reliability/availability offered by the transmission alternatives is necessary to make a planning selection between either ac or dc transmission.

1.2.1 Evaluation Of Transmission Costs

The cost of a transmission line comprises of the capital investment required for the actual infrastructure (i.e. Right of Way (RoW), towers, conductors, insulators and terminal equipment) and costs incurred for operational requirements (i.e. losses). Assuming similar insulation requirements for peak voltage levels for both ac and dc lines, a dc line can carry as much power with two conductors (having positive/negative polarities with respect to ground) as an ac line with three conductors of the same size. Therefore for a given power level, a dc line requires smaller RoW, simpler and cheaper towers and reduced conductor and insulator costs. As an example, Figure 1-1, shows the comparative case of ac and dc systems carrying 2000 MW.

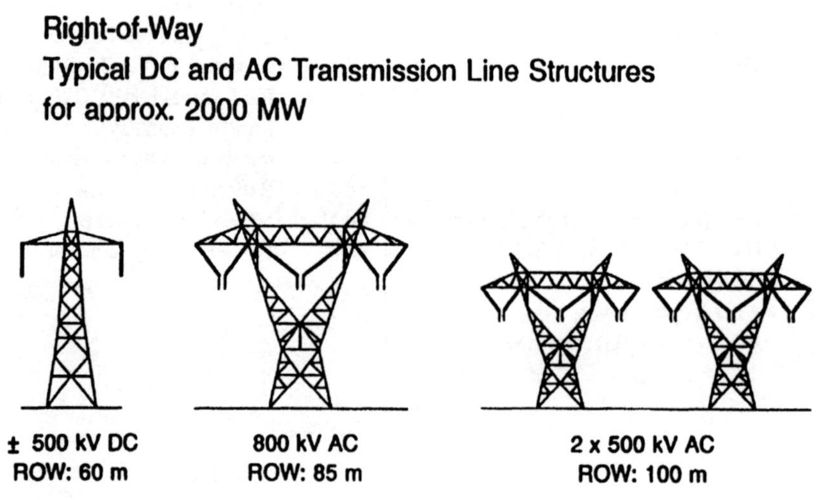

Figure 1-1: Comparison of RoW for ac-dc transmission systems

With the dc option, since there are only two conductors (with the same current capacity of 3 ac conductors), the power transmission losses are also reduced to about two-thirds of the comparable ac system. The absence of skin effect with dc is also beneficial in reducing power losses marginally, and the dielectric losses in case of power cables is also very much less for dc transmission.

Corona effects tend to be less significant on dc than for ac conductors. The other factors that influence line costs are the costs of compensation and terminal equipment. dc lines do not require reactive power compensation but the terminal equipment costs are increased due to the presence of converters and filters.

Figure 1-2 shows that variation of costs of transmission with distance for ac and dc transmission. AC tends to be more economical than dc for distances less than the "breakeven distance" but is more expensive for longer distances. The breakeven distances can vary between 400 to 700 km in overhead lines depending on the per unit line costs. With a cable system, this breakeven distance lies between 25 to 50 km.

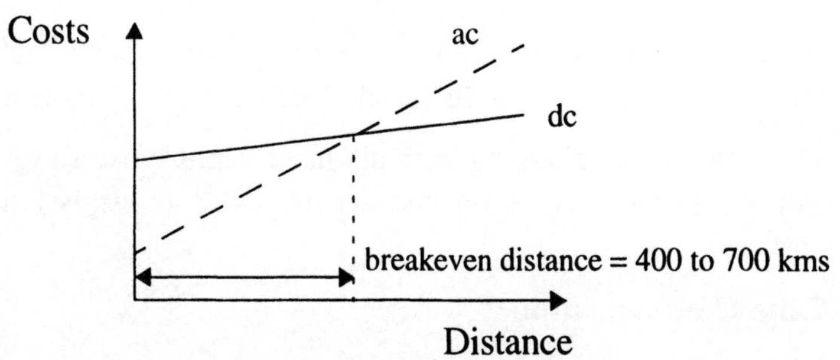

Figure 1-2: Comparison of ac/dc lines

1.2.2 Evaluation Of Technical Considerations

Due to its fast controllability, a dc transmission has full control over transmitted power, an ability to enhance transient and dynamic stability in associated ac networks and can limit fault currents in the dc lines. Furthermore, dc transmission overcomes some of the following problems associated with ac transmission:

- **Stability Limits**

  The power transfer in an ac line is dependent on the angle difference between the voltage phasors at the two line ends. For a given power transfer level, this angle increases with distance. The maximum power transfer is limited by the considerations of steady state and transient stability. The power carrying capability of an ac line is inversely proportional to transmission distance whereas the power carrying ability of dc lines is unaffected by the distance of transmission.

- **Voltage Control**

Voltage control in ac lines is complicated by line charging and voltage drops. The voltage profile in an ac line is relatively flat only for a fixed level of power transfer corresponding to its Surge Impedance Loading (SIL). The voltage profile varies with the line loading. For constant voltage at the line ends, the midpoint voltage is reduced for line loadings higher than SIL and increased for loadings less than SIL.

The maintenance of constant voltage at the two ends requires reactive power control as the line loading is increased. The reactive power requirements increase with line length.

Although dc converter stations require reactive power related to the power transmitted, the dc line itself does not require any reactive power.

The steady-state charging currents in ac cables pose serious problems and makes the break-even distance for cable transmission around 50 kms.

- **Line Compensation**

Line compensation is necessary for long distance ac transmission to overcome the problems of line charging and stability limitations. The increase in power transfer and voltage control is possible through the use of shunt inductors, series capacitors, Static Var Compensators (SVCs) and, lately, the new generation Static Compensators (STATCOMs).

In the case of dc lines, such compensation is not needed.

- **Problems of AC Interconnection**

The interconnection of two power systems through ac ties requires the automatic generation controllers of both systems to be coordinated using tie line power and frequency signals. Even with coordinated control of interconnected systems, the operation of ac ties can be problematic due to:

1. The presence of large power oscillations which can lead to frequent tripping,

2. Increase in fault level, and

3. Transmission of disturbances from one system to the other.

The fast controllability of power flow in dc lines eliminates all of the above problems. Furthermore, the asynchronous interconnection of two power systems can only be achieved with the use of dc links.

- **Ground Impedance**

In ac transmission, the existence of ground (zero sequence) current cannot be permitted in steady-state due to the high magnitude of ground impedance which will not only affect efficient power transfer, but also result in telephonic interference.

The ground impedance is negligible for dc currents and a dc link can operate using one conductor with ground return (monopolar operation). The ground return is objectionable only when buried metallic structures (such as pipes) are present and are subject to corrosion with dc current flow. It is to be noted that even while operating in the monopolar mode, the ac network feeding the dc converter station operates with balanced voltages and currents. Hence, single pole operation of dc transmission systems is possible for extended period, while in ac transmission, single phase operation (or any) unbalanced operation) is not feasible for more than a second.

- **Problems of DC Transmission**

The application of dc transmission is limited by factors such as:

1. High cost of conversion equipment,
2. Inability to use transformers to alter voltage levels,
3. Generation of harmonics,
4. Requirement of reactive power, and
5. Complexity of controls.

Over the years, there have been significant advances in dc technology, which have tried to overcome the disadvantages listed above, except for item (2). These advances in dc technology are:

1. Increase in the ratings of a thyristor cell that makes up a valve,

2. Modular construction of thyristor valves,
3. Twelve-pulse (and higher) operation of converters,
4. Use of forced-commutation [4], and
5. Application of digital electronics and fiber optics in the control of converters.

Some of the above advances have resulted in improving the reliability and reduction of conversion costs in dc systems.

1.2.3 Evaluation Of Reliability And Availability Costs

Statistics on the reliability of HVDC links are maintained by CIGRE and IEEE Working Groups. The reliability of dc links has been very good and is comparable with ac systems. The availability of dc links is quoted in the upper 90%.

1.2.4 Applications of DC Transmission

Due to their costs and special nature, most applications of dc transmission generally fall into one of the following four categories:

- **Underground or underwater cables**

  In the case of long cable connections over the breakeven distance of about 40-50 km, dc cable transmission system has a marked advantage over ac cable connections. Examples of this type of applications were the Gotland (1954) and Sardinia (1967) schemes.

  The recent development of Voltage Source Converters (VSC) and the use of rugged polymer dc cables, with the so-called "HVDC Light" option, is being increasingly considered. An example of this type of application is the 180 MW Directlink connection (2000) in Australia.

- **Long distance bulk power transmission**

  Bulk power transmission over long distances is an application ideally suited for dc transmission and is more economical than ac transmission whenever the breakeven distance is exceeded. Examples of this type of application abound from the earlier Pacific Intertie to the recent links in China and India.

  The breakeven distance is being effectively decreased with the reduced costs of new compact converter stations possible due to the recent advances in power electronics (discussed in a later section).

- **Asynchronous interconnection of ac systems**

  In terms of an asynchronous interconnection between two ac systems, the dc option reigns supreme. There are many instances of BB connections where two ac networks have been tied together for the overall advantage to both ac systems. With recent advances in control techniques, these interconnections are being increasingly made at weak ac systems. The growth of BB interconnections is best illustrated with the example of N.America where the four main independent power systems are interconnected with twelve BB links.

  In the future, it is anticipated that these BB connections will also be made with VSCs offering the possibility of full four-quadrant operation and the total control of active/reactive power coupled with the minimal generation of harmonics.

- **Stabilization of power flows in integrated power system**

  In large interconnected systems, power flow in ac ties (particularly under disturbance conditions) can be uncontrolled and lead to overloads and stability problems thus endangering system security. Strategically placed dc lines can overcome this problem due to the fast controllability of dc power and provide much needed damping and timely overload capability. The planning of dc transmission in such applications requires detailed study to evaluate the benefits. Examples are the IPP link in the USA and the Chandrapur-Padghe link in India.

Presently the number of dc lines in a power grid is very small compared to the number of ac lines. This indicates that dc transmission is justified only for specific applications. Although advances in technology and introduction of Multi-Terminal DC (MTDC) systems are expected to increase the scope of application of dc transmission, it is not anticipated that the ac grid will be replaced by a dc power grid in the future. There are two major reasons for this:

- First, the control and protection of MTDC systems is complex and the inability of voltage transformation in dc networks imposes economic penalties.

- Second, the advances in power electronics technology have resulted in the improvement of the performance of ac transmissions using FACTS devices, for instance through introduction of static var systems, static phase shifters, etc.

1.3 TYPES OF HVDC SYSTEMS

Three types of dc links are considered in HVDC applications.

1.3.1 Monopolar Link

A monopolar link (Figure 1-3a) has one conductor and uses either ground and/or sea return. A metallic return can also be used where concerns for harmonic interference and/or corrosion exist. In applications with dc cables (i.e. HVDC Light), a cable return is used. Since the corona effects in a dc line are substantially less with negative polarity of the conductor as compared to the positive polarity, a monopolar link is normally operated with negative polarity.

1.3.2 Bipolar Link

A bipolar link (Figure 1-3b) has two conductors, one positive and the other negative. Each terminal has two sets of converters of equal rating, in series on the dc side. The junction between the two sets of converters is grounded at one or both ends by the use of a short electrode line. Since both poles operate with equal currents under normal operation, there is zero ground current flowing under these conditions. Monopolar operation can also be used in the first stages of the development of a bipolar link. Alternatively, under faulty converter conditions, one dc line may be temporarily used as a metallic return with the use of suitable switching.

1.3.3 Homopolar Link

In this type of link (Figure 1-3c) two conductors having the same polarity (usually negative) can be operated with ground or metallic return.

Due to the undesirability of operating a dc link with ground return, bipolar links are mostly used. A homopolar link has the advantage of reduced insulation costs, but the disadvantages of earth return outweigh the advantages.

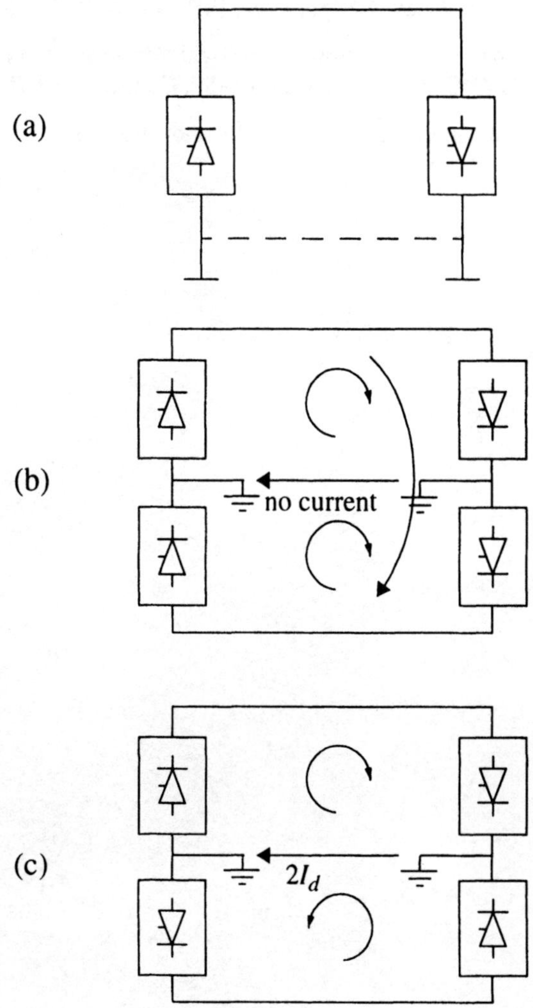

Figure 1-3: Types of HVDC links (a) monopolar link (b) bipolar link and (c) homopolar dc link

1.4 REFERENCES

[1]. Direct Current Transmission - Volume I, E.W. Kimbark, Wiley Interscience, 1971, ISBN 0-471-35550-X.

[2]. High Voltage Direct Current Transmission, 2nd Edition, The Institution of Electrical Engineers, J. Arrillaga, 1998, ISBN 0-85296-941-4.

[3]. HVDC Power Transmission Systems - Technology and System Interactions, K.R. Padiyar, John Wiley & Sons, 1990, ISBN 0-470-21706-5.

[4]. V.K.Sood, Position paper for Canadian Electrical Association on "Artificially Commutated HVDC Inverters", March 1989, Contract No. ST-174B.

[5]. HVDC Projects Listing, Prepared by IEEE DC and Flexible AC Transmission Subcommittee, D. Melvold.

Chapter 2

Types of Converters

2.1 INTRODUCTION

A HVDC system requires an electronic converter for its ability of converting electrical energy from ac-dc or vice versa. There are basically two configuration types of three-phase converters possible for this conversion process (Figure 2-1):

- Current Source Converter (CSC), and
- Voltage Source Converter (VSC).

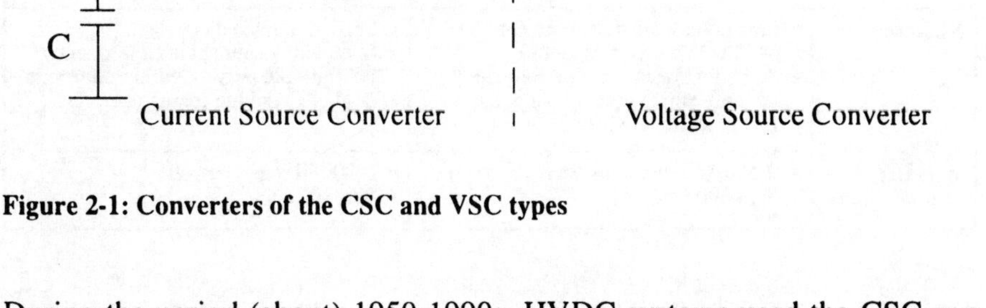

Figure 2-1: Converters of the CSC and VSC types

During the period (about) 1950-1990s, HVDC systems used the CSC configuration almost exclusively. The traditional CSC utilized the mercury-arc valve from the early 1950s to the mid-1970s, and thereafter, the thyristor valve as its fundamental switching device.

From about 1990 onwards, the alternative VSC became economically viable due to the availability of new self-commutating high-power switches (such as GTOs and IGBTs) and the computing power of DSPs to generate the appropriate firing patterns.

Modern HVDC transmission systems can utilize either the traditional Current Source Converter (CSC) or the Voltage Source Converter (VSC) as the basic conversion workhorse. The two converters are actually duals of one another. However, the choice of which option is selected for a particular project is based upon economic and other factors. A comparison of the characteristics of the two converter types is made in Table 2-1. However, at present VSC are still limited to below 250 MW capacity due to commercial and practical limitations of the electronic switches.

Table 2-1: Comparison of converter types

	Converter type	
	CSC	VSC
On AC side	• Acts as a constant voltage source • Requires a capacitor as its energy storing device • Requires large ac filters for harmonic elimination • Requires reactive power supply for power factor correction	• Acts as a constant current source • Requires an inductor as its energy storing device • Requires only a small ac filter for higher harmonics elimination • Reactive power supply is not required as converter can operate in any quadrant
On DC side	• Acts as a constant current source • Requires an inductor as its energy storing device • Requires dc filters • Provides inherent fault current limiting features	• Acts as a constant voltage source • Requires a capacitor as its energy storing device • Energy storage capacitor provide dc filtering capability at no extra cost • Problematic for dc line side faults since the charged capacitor will discharge into the fault
Switches	• Line commutated or force commutated with a series capacitor • Switching occurs at line frequency i.e. only single pulsing per cycle • Lower switching losses	• Self-commutated • Switching occurs at high frequency i.e. multiple pulsing within one cycle • Higher switching losses
Rating range	• 0-550 MW per converter • upto 600 kV	• 0-200 MW per converter • upto100 kV

This chapter will present a simplified theory and approach of the operational behavior of the two types of converters.

2.2 CURRENT SOURCE CONVERTERS (CSC)

Since the basics of 6-pulse bridge CSC converter theory are already well documented in other books [1], this chapter will be kept brief and is included for the sake of completeness only so that the book is self-contained. Moreover, the terminology that will be employed in later chapters will be introduced here.

To consider the theoretical analysis of a conventional 6-pulse bridge (Figure 2-2), the following assumptions are made:

- DC current I_d is constant (i.e. the smoothing reactor L_d is infinite),
- Valves are ideal switches, and
- AC system is infinitely strong (i.e. the 3 phase emfs are balanced and perfectly sinusoidal).

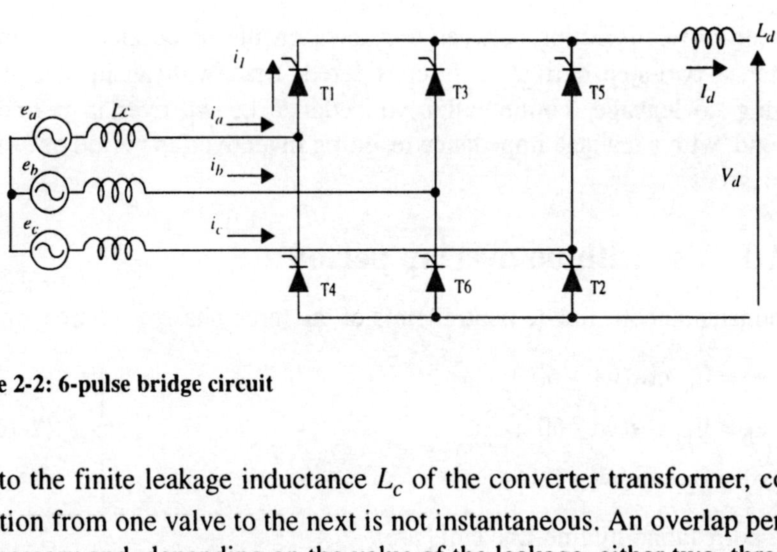

Figure 2-2: 6-pulse bridge circuit

Due to the finite leakage inductance L_c of the converter transformer, commutation from one valve to the next is not instantaneous. An overlap period is necessary and, depending on the value of the leakage, either two, three or four valves may conduct at any one time. In the most general case, with a typical value of converter transformer leakage impedance of between 13-18%, either two or three valves conduct at any one time. Essentially, this results as an overlap angle μ which is less than 60 degrees; typically, it is in the range of 20 to 25 degrees. During commutation, three valves conduct and in-between commutations only two valves conduct. The case of either two or three conducting valves is shown in Figure 2-3.

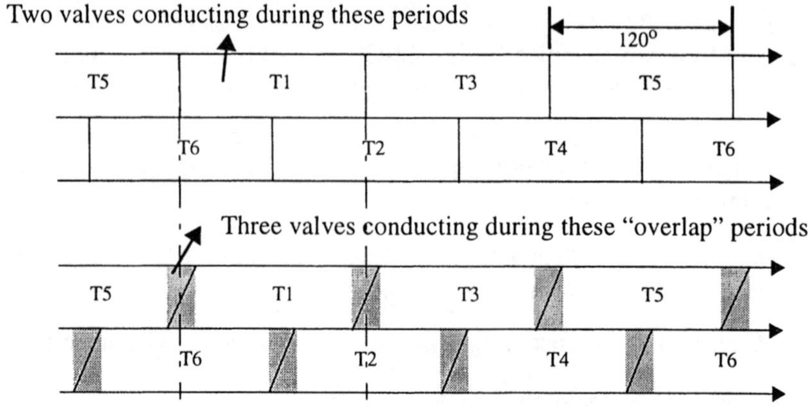

Figure 2-3: Effect of overlap on the number of conducting valves

To derive relationships between variables on the ac-dc sides of the converter, a two-step derivation is considered: first, with an ideal converter having no leakage (commutation) impedance i.e. no overlap period, and second, with a leakage impedance resulting in an overlap period of less than 60 degrees.

2.2.1 Case with no overlap period

The instantaneous line to neutral emfs of the three phase ac source are:

$$\left.\begin{aligned} e_a &= E_m \cos(wt + 60^o) \\ e_b &= E_m \cos(wt - 60^o) \\ e_c &= E_m \cos(wt - 180^o) \end{aligned}\right\} \quad (2\text{-}1)$$

The corresponding line-line emfs are:

$$\left.\begin{aligned} e_{ac} &= e_a - e_c = 1.732{*}E_m \cos(wt + 30^o) \\ e_{ba} &= e_b - e_a = 1.732{*}E_m \cos(wt - 90^o) = 1.732{*}E_m \sin(wt) \\ e_{cb} &= e_c - e_b = 1.732{*}E_m \cos(wt + 150^o) \end{aligned}\right\} (2\text{-}2)$$

These waveforms are shown in Figure 2-4.

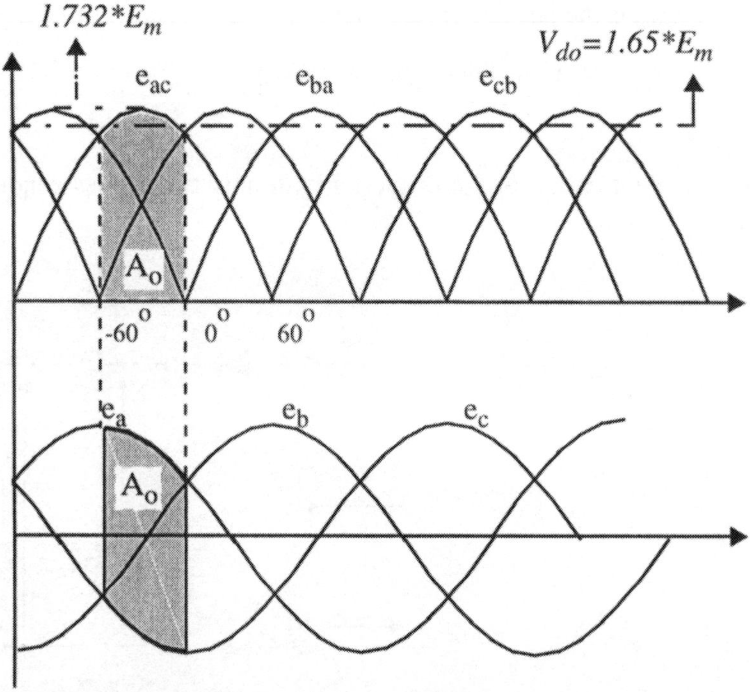

Figure 2-4: Waveforms of line-to-line and phase voltages

The dc voltage across the bridge is the difference between the line to phase voltages, such as the shaded area A_o (Figure 2-4) during which time valves T1 and T5 are conducting.

The average direct voltage V_d is found by integrating the instantaneous voltages over a 60 degree period:

$$V_{do} = \frac{3}{\pi}\int_{-60^o}^{0} e_{ac}d\theta = \frac{3}{\pi}\int_{-60^o}^{0} E_m\cos(\theta + 30^o)d\theta \qquad (2\text{-}3)$$

$$V_{do} = 1.65E_m = 1.35E_{LL} \qquad (2\text{-}4)$$

where V_{do} is called the ideal no-load direct voltage with $\alpha = 0$ degs.,

E_m is the peak ac voltage, and

E_{LL} is the rms line-to-line ac voltage.

The equivalent circuit of the converter with only two valves conducting is shown in Figure 2-5.

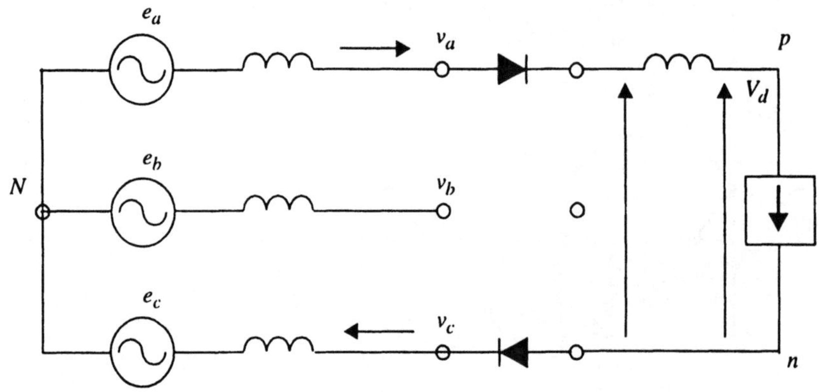

Figure 2-5: Bridge converter with two valves conducting

With a delay angle of α, the limits of integration are simply increased by α, and therefore

$$V_d = V_{do}\int_{(\alpha - 60^0)}^{\alpha} \cos(\theta + 30^o)d\theta \tag{2–5}$$

$$V_d = V_{do}\cos\alpha \tag{2–6}$$

Relationship between ac and dc current

Assuming no losses in the bridge, the ac power fed into the bridge is equal to the dc power emanating from the bridge i.e.

$$P_d = V_d \cdot I_d = 3\, E_{LN} \cdot I_{L1} \cdot \cos\phi \tag{2–7}$$

where I_{L1} is the fundamental frequency component of the line current and ϕ is the phase angle between the line voltage and line current. The line voltages and line current are depicted in Figure 2-6. The line current has a rectangular shape, and with no overlap i.e. no delay angle, it will be in phase with the line voltage. However, with a firing delay, this current will be displaced by an angle α.

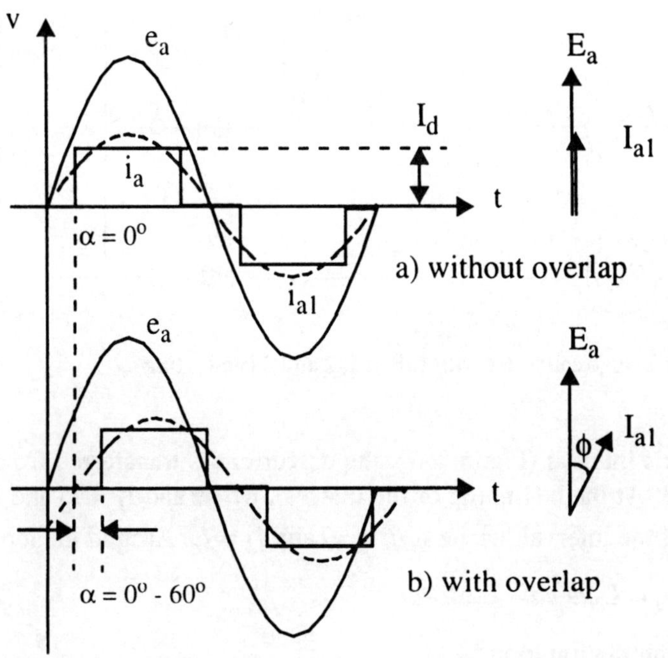

Figure 2-6: Relationship between ac and dc currents with and without overlap

By Fourier Analysis, the value of the line current is given by

$$I_{L1} = \frac{\sqrt{2}}{\pi} \cdot \int_{-60^o}^{60^0} (I_d \cdot \cos\theta) d\theta \tag{2-8}$$

$$I_{L1} = \frac{\sqrt{6}}{\pi} \cdot I_d \tag{2-9}$$

2.2.2 Case with overlap period less than 60 degrees.

With reference to the condition with commutation impedance and overlap period less than 60 degrees, i.e. when three valves conduct,

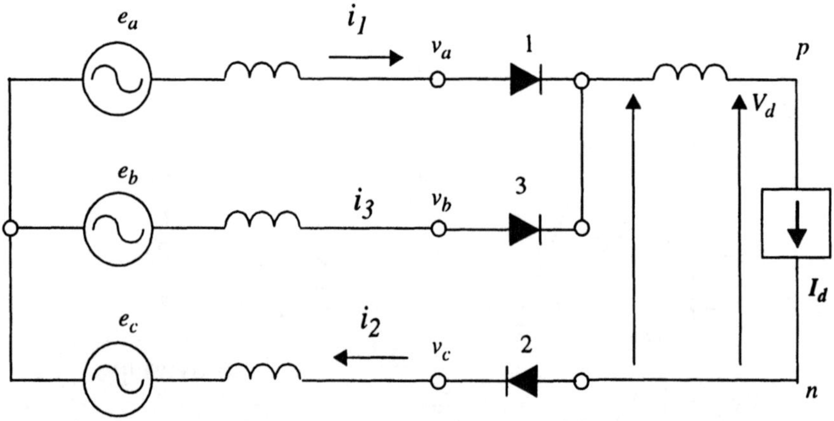

Figure 2-7: Bridge converter with valves 1, 2 and 3 conducting

During this interval (Figure 2-7), the dc current is transferred from valve 1 to valve 3. At the beginning of the interval, $wt=\alpha$ and $i_1 = I_d$ and $i_3= 0$. At the end of the interval, $wt=\alpha + \mu$, $i_1=0$ and $i_3 = I_d$. Around the loop,

$$e_b - e_a = L_c.di_3/dt - L_c.di_1/dt \quad (2\text{-}10)$$

In the commutation loop,

$$e_b - e_a = 1.732.\ E_{LL}.\ sin\ wt \quad (2\text{-}11)$$

and

$$i_1 = I_d - i_3 \quad (2\text{-}12)$$

Therefore, $di_1/dt = 0 - di_3/dt$ (2-13)

We can consider the emf in the commutation loop as

$$1.732.\ E_{LL}\ sin\ wt = 2\ L_c\ di_3/dt \quad (2\text{-}14)$$

Integrating this equation gives

$$i_3 = 1.732.\ E_{LL}/2wL_c\ .(cos\ \alpha - cos\ wt) \quad (2\text{-}15)$$

This equation shows that the current i_3 is composed of a dc component and a sinusoidal component. The sinusoidal component lags the commutation voltage by 90 degs. since the loop is purely inductive, and it has a peak value of $1.732.\ E_{LL}/2wL_c$. This current value is that of a line to line short circuit on the ac source. The dc component depends on the term cos α.

At the end of commutation, $wt = \alpha + \mu$, and $i_3 = I_d$. Using this in eq. (8) gives

$$I_d = 1.732.\ V_{LL}/2wL_c\ .(\cos\alpha - \cos(\alpha + \mu)) \qquad (2\text{-}16)$$

The average dc voltage of the converter (represented by area A_o in the Figure 2-8) is reduced due to the effect of overlap caused by a voltage drop ΔV_d, represented by the area A in Figure 2-8, across the commutation impedance. During commutation, the line-to-line voltage of the short circuited phase is zero. This implies that the line-to-neutral (phase) voltages are equal to each other, and that the two phase impedances jwL_c act as a voltage divider.

The dc voltage drop, represented by area A, is given by:

$$A = \int_{\alpha}^{(\alpha+\mu)} \left(e_b - \frac{e_a + e_b}{2}\right) d\theta \qquad (2\text{-}17)$$

$$A = 0.866 V_m \cdot (\cos\alpha - \cos(\alpha + \mu)) \qquad (2\text{-}18)$$

The average voltage drop ΔV_d is given by:

$$\Delta V_d = \frac{3}{\pi} A = \frac{3}{\pi} \cdot 0.866 V_m \cdot (\cos\alpha - \cos(\alpha + \mu)) \qquad (2\text{-}19)$$

Since $I_d = 1.732.\ V_{LL}/2wL_c\ .(\cos\alpha - \cos(\alpha + \mu))$, from eq. 2-16, dividing these two eqs. and using the relationship between V_m and V_{LL} gives the relationship between the voltage drop and the dc current.

Therefore, the dc voltage with overlap is given by:

$$V_d = V_{do} \cdot \cos\alpha - \Delta V_d \tag{2-20}$$

where $\Delta V_d = \frac{3}{\pi} \cdot wL_c \cdot I_d$

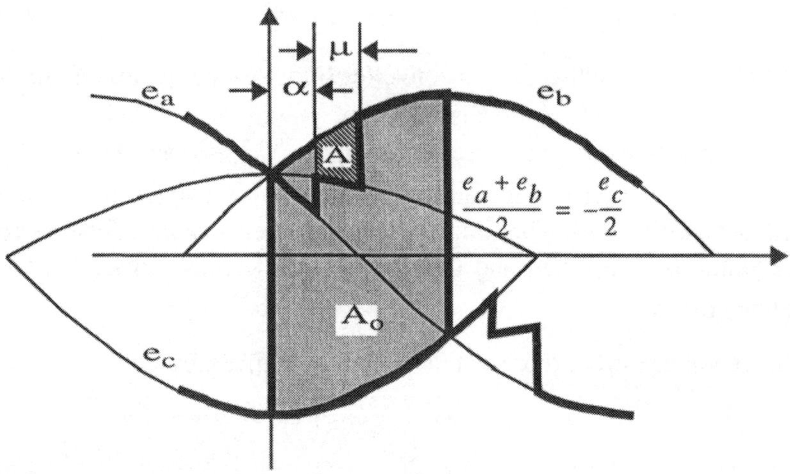

Figure 2-8: DC voltage with the effects of overlap

For a rectifier, the analysis of the bridge gives the following dc output voltages:

$$V_{dr} = V_{dor} \cdot \cos\alpha - R_{cr} \cdot I_d \tag{2-21}$$

where

$V_{dor} = \frac{3}{\pi} \cdot \sqrt{2} \cdot V_{LL}$ and $R_{cr} = \frac{3}{\pi} \cdot wL_{cr}$

Results from an EMTP simulation of a 6 pulse rectifier are shown in Figure 2-9. Shown are, from top to bottom, the three phase ac voltages, two valve currents, two valve voltages, dc current, dc voltage and commutation voltages on the secondary of the converter transformer. Seen clearly are the overlap and firing angles, as well as the harmonic ripple on the signals of dc current and dc voltage. Also seen are the commutation notches on the ac voltages at the transformer secondary.

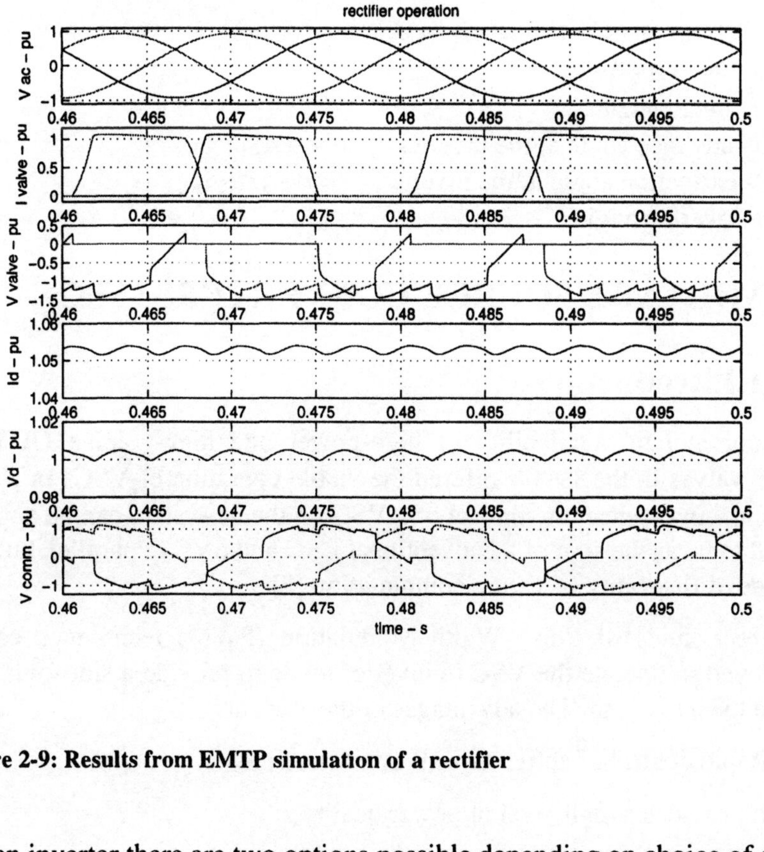

Figure 2-9: Results from EMTP simulation of a rectifier

For an inverter there are two options possible depending on choice of either the delay angle (eq.(2)) or extinction angle (eq.(3)) as the control variable.

$$-V_{di} = V_{doi} \cdot \cos\beta + R_{ci} \cdot I_d \tag{2-22}$$

$$-V_{di} = V_{doi} \cdot \cos\gamma - R_{ci} \cdot I_d \tag{2-23}$$

where:

V_{dr} and V_{di} - dc voltage at the rectifier and inverter respectively
V_{dor} and V_{doi} - open circuit dc voltage at the rectifier and inverter respectively
R_{cr} and R_{ci} - equivalent commutation resistance at the rectifier and inverter respectively
L_{cr} and L_{ci} - leakage inductance of converter transformer at rectifier and inverter respectively
I_d - dc current
α - delay angle
β - advance angle at the inverter, ($\beta = \pi - \alpha$)
γ - extinction angle at the inverter, ($\gamma = \pi - \alpha - \mu$)
μ - overlap angle

2.3 VOLTAGE SOURCE CONVERTERS (VSC)

2.3.1 Introduction

The commercial availability of high-power and high-voltage GTO and IGBT valves in the 1990's offered the viable operation of VSCs in HVDC schemes. In essence, the impact of a VSC on the ac system can be approximated to be as the sum of a conventional CSC and SVC in parallel, but with the added flexibility of secure commutation [2].

Different kinds of Pulse Width Modulation (PWM) techniques can be employed to operate the VSC in inverter mode to provide a sinusoidal output to the ac system. The advantages of the VSC are:

- Rapid control of active as well as reactive power,
- It provides a high level of power quality,
- Minimal environmental impact, and
- Ability to connect to weak ac networks, or even dead networks.

The technology lends itself to the following types of applications:

- Low power (less than 250 MW) HVDC transmission (commercially referred to as "HVDC Light"),
- VAR Computation (SVC and STATCOM), and
- Active Filters.

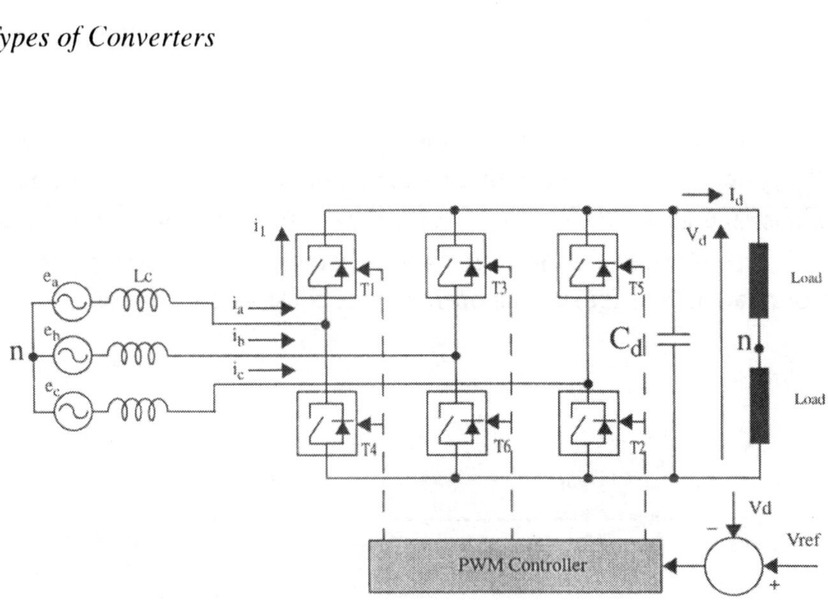

Figure 2-10: VSC operating principles [2]

VSCs utilize self-commutating switches (e.g. GTOs, IGBTs) which can be turned-on or off at will. This is in contrast to the conventional CSCs which operate with line-commutated thyristor switches. Commutation in a force-commutated VSC valve can occur many times per cycle, whereas in a line-commutated CSC it can happen only once per cycle. This feature allows the voltage/current in a VSC to be modulated to produce a nearly sinusoidal output and control the power factor as well. Furthermore, power reversal in a VSC can be made with either current or voltage reversal at the dc side. In contrast, with a CSC, power reversal can occur only with voltage reversal.

In Figure 2-10, the operating principles of a VSC are evident. The dc side capacitor C_d and ac side inductor L_c are necessary elements of the VSC. The dc voltage V_d is monitored and compared to a reference value V_{ref} to generate an error signal which controls the PWM controller. When the dc current I_d is positive, the VSC acts as a rectifier; the dc capacitor is discharged as it feeds the dc load, and the control system will modify the firing angle to import power from the ac system. When the dc current I_d is negative, the VSC acts as an inverter; the dc capacitor is charged from the dc source, and the control system will modify the firing angle to export power to the ac system.

The VSC can also modulate the firing of the valves to control the reactive power so that a unity power factor (or any other value, for that matter) can be obtained.

The PWM controller generates a voltage V_{gen} with the same frequency as the ac system voltage V_s. By altering the amplitude of V_{gen} and its phasor relationship with V_s, the converter can be made to operate in all four quadrants i.e. rectifier/inverter operation with lagging/leading power factor. The phasor relationships for such an operation are illustrated in Figure 2-11.

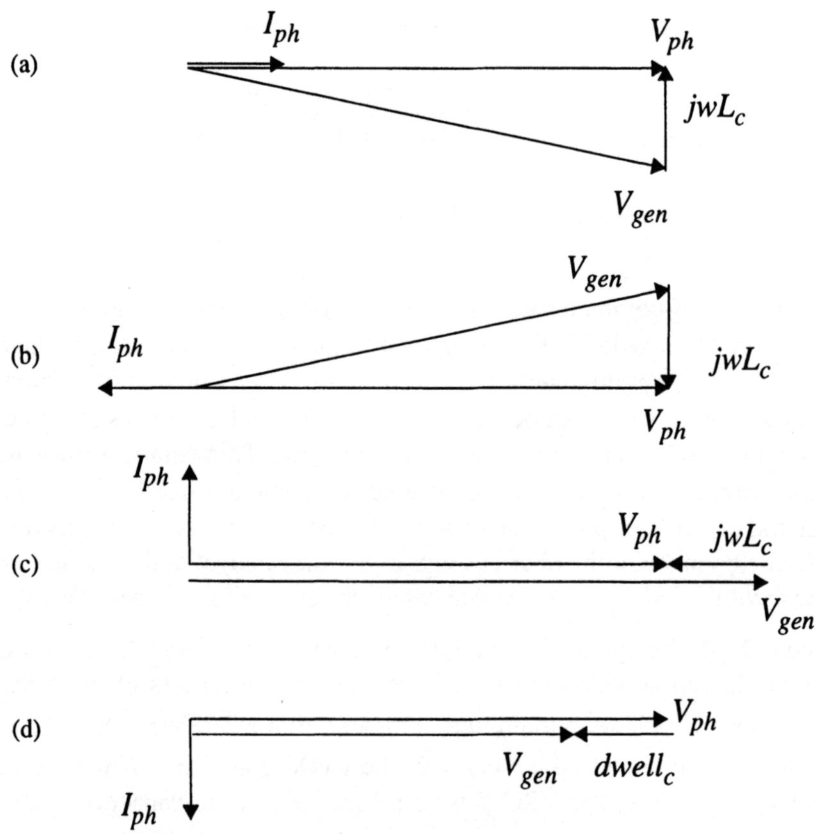

Figure 2-11: Four quadrant operation of the VSC

(a) rectifier operation at unity power factor,

(b) inverter operation at unity power factor,

(c) purely reactive with leading current, and

(d) purely reactive with lagging current

Figure 2-12 illustrates the operation of a VSC inverter feeding an inductive load from a dc supply. Figure 2-12a shows the output voltage generated by the inverter and its fundamental component. Figure 2-12b shows the output current before filtering. The PWM modulation is clearly visible. Figure 2-12c shows the valve (switch with anti-parallel diode) current; the positive half is current in the switch whereas the negative half is the diode current. And Figure 2-12d shows the voltage across the valve with the PWM pattern.

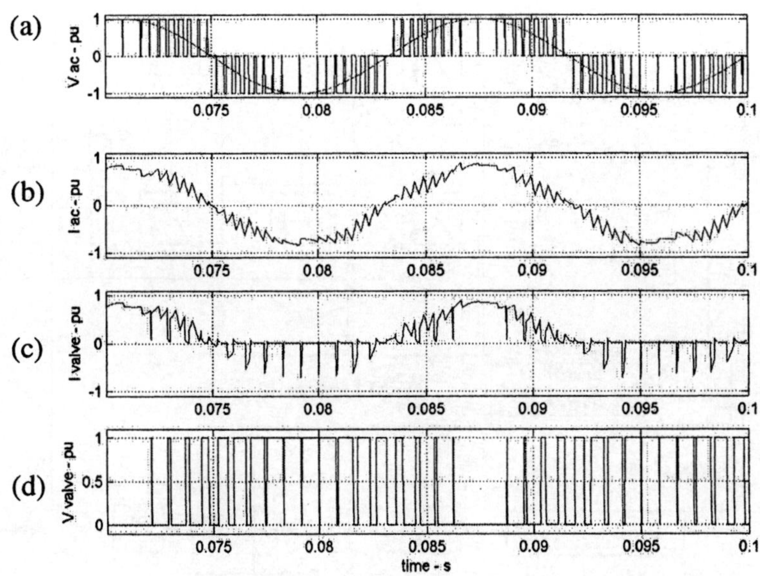

Figure 2-12: Results from a VSC operation

2.3.2 Control of the DC Capacitor Voltage

As explained earlier, the dc capacitor voltage of the VSC can be modulated by a PWM controller as a function of either ac current or voltage. The version with ac current control is simpler and more stable than the version with ac voltage control. These two versions will be described next.

2.3.3 VSC with AC Current Control

The VSC with ac current control method is shown in Figure 2-13. The objective here is to force the phase currents to follow a sinusoidal current reference template I_{ref}. The amplitude of the current reference template is made a function of the dc capacitor voltage V_d and the desired dc voltage V_{ref} i.e.

$$I_{max} = K(V_{ref} - V_d) \tag{2-24}$$

where K is the gain of a PI controller, and equal to $K_p + K_i/s$.

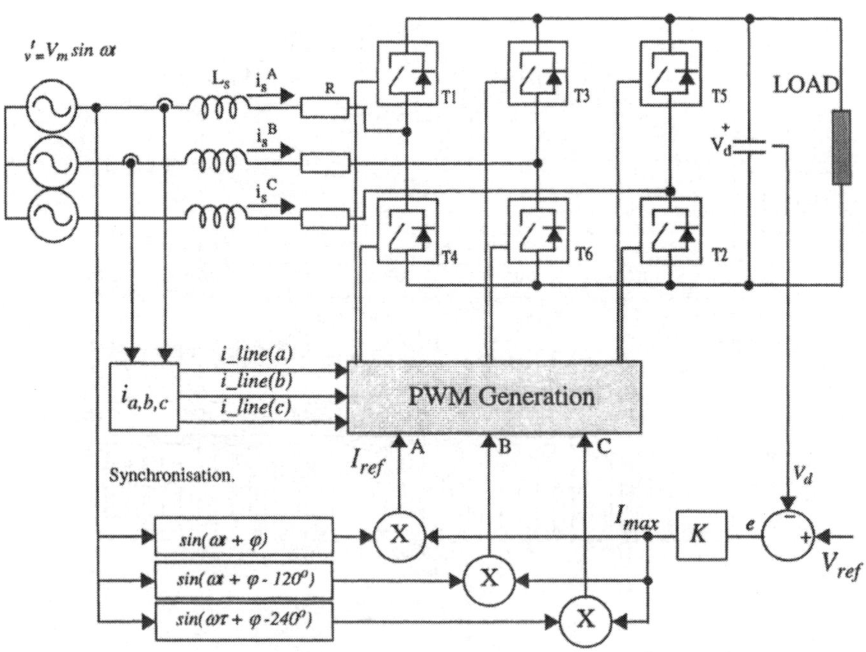

Figure 2-13: Voltage-source current-controlled PWM rectifier [2]

The desired sinusoidal current waveform template is obtained as

$$I_{ref} = I_{max} \sin(wt + \varphi), \tag{2-25}$$

where w is the same as the supply frequency and φ is the desired phase shift. Therefore, synchronization with the supply voltage(s) is necessary to obtain a correct version of the $sin(wt+\varphi)$. A measure of the instantaneous phase currents is necessary to generate an error between them and the reference current template, and this error is minimized by means of the closed feedback loop.

In a VSC with ac current control, therefore, it is necessary to monitor the three phase ac voltages and currents (note that, as shown in the figure, a two-phase current measurement is acceptable if there is no neutral connection).

Next, a modulation technique is necessary to generate the PWM pattern for firing of the converter switches such that the current follows the reference template value. Different techniques are feasible, and these are described next.

2.3.3.1 PWM Pattern Generation Techniques

Three widely used PWM modulation techniques for following the current reference template are described here:

1. Periodical Sampling (PS) (Figure 2-14)

This method employs a clock of fixed frequency to control the power switches of the VSC. An error signal, generated from a measurement of the line and reference currents, is used to modulate the clock frequency and generate the PWM pattern using a comparator and D-type flip-flop. The minimum time between switching is limited by the clock frequency.

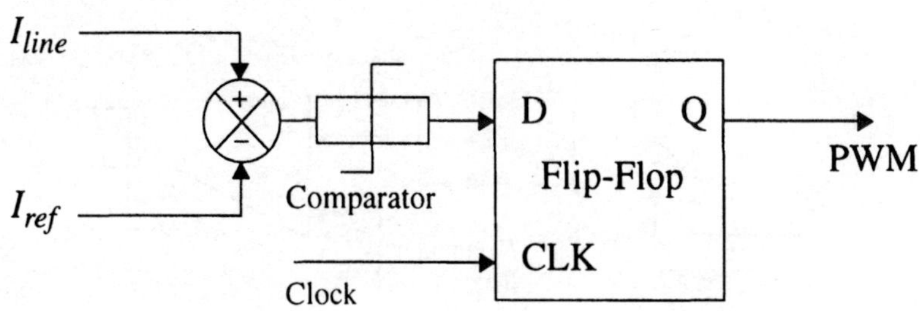

Figure 2-14: Periodic Sampling technique for PWM pattern generation

2. Hysteresis Band (HB) (Figure 2-15)

The Hysteresis Band (HB) technique modulates the converter switches when the error between the I_{line} and I_{ref} exceeds by a fixed magnitude of I_{ref} (typically 5-10% of the set value). In this case, the switching frequency is not fixed and varies as a function of the I_{ref} magnitude, the hysteresis band h and the inductance in the load.

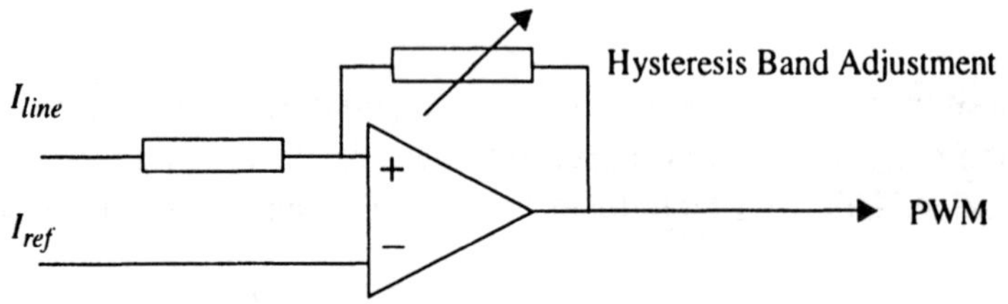

Figure 2-15: Hysteresis Band (HB) technique for PWM pattern generation

3. Triangular Carrier (TC) Technique (Figure 2-16)

The TC method compares the error between the I_{line} and I_{ref} with a fixed-frequency, fixed-amplitude triangular carrier wave. A PI regulator provides the static and dynamic properties to the feedback loop. This method is more complex than the other two methods in its implementation since the gains of the PI controller need to be selected.

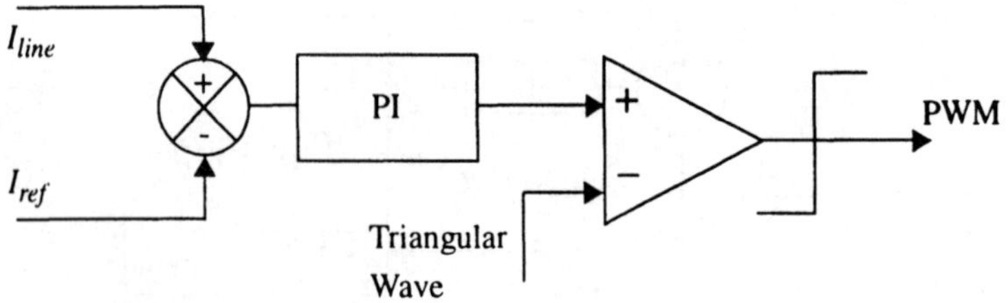

Figure 2-16: Triangular carrier wave technique for PWM pattern generation

2.3.4 VSC with AC Voltage Control

Figure 2-17 shows an equivalent single-phase version of the VSC with voltage control. In this circuit, the generated voltage V_{gen} is able to be controlled both in amplitude and phase with the source voltage V_s. An advantage of this method is that this method does not require the input ac currents to be monitored. The template for the V_{gen} is obtained below.

From the figure,

$$v_s = L_s\, di/dt\, R_{is} + V_{gen} \quad (2\text{-}25)$$

Assuming that $v_s = V*sqrt(2)$, then the solution for the current should be of the form

$$i_s = I_{max}(t).sin(wt + f) \quad (2\text{-}26)$$

Solving for V_{gen} gives

$$V_{gen} = \left[V \cdot \sqrt{2} + X_s I_{max} \sin\varphi - \left(RI_{max} + L_s \cdot \frac{dI_{max}}{dt}\right)\right] \cdot \sin wt$$
$$-\left[X_s I_{max} \cdot \cos\varphi + \left(RI_{max} + L_s \cdot \frac{dI_{max}}{dt}\right) \cdot \sin\varphi\right] \cdot \cos wt$$

(2-27)

This equation provides a means to control the V_{gen} through modulation of I_{max}. This equation also controls the power factor $cos\varphi$. In case unity power factor is desired then $cos\varphi = 1$ and $sin\varphi = 0$, and can be used to simplify the above equation.

$$V_{gen} = \left[V \cdot \sqrt{2} - \left(RI_{max} + L_s \cdot \frac{dI_{max}}{dt}\right)\right] \cdot \sin wt \quad -[X_s I_{max}] \cdot \cos wt$$

(2-28)

This algorithm is solved with the implementation shown in Figure 2-17. Note that the equation above has an in-phase and in-quadrature terms which allow it to modulate both the amplitude and phase of the generated voltage V_{gen}.

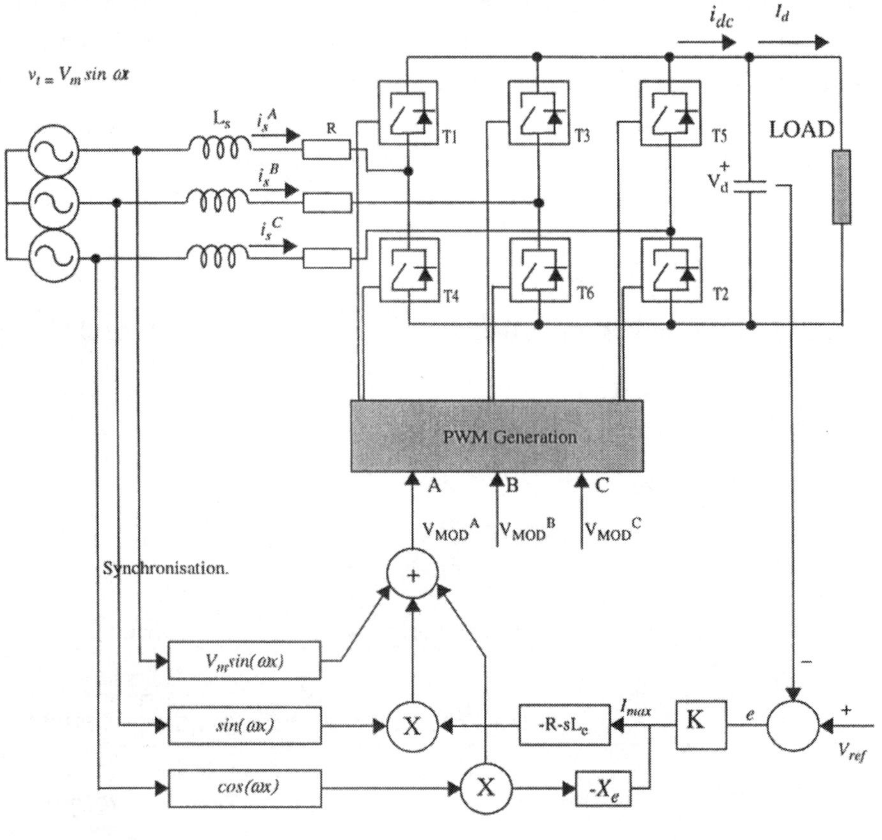

Figure 2-17: Implementation of the voltage-controlled rectifier for unity power factor operation [2]

When compared to the VSC with current control, there is no need to monitor the three phase currents to implement this algorithm. However, due to the presence of the terms R and X_s, an exact knowledge of the source impedance is necessary. This, in a practical implementation, is highly unlikely as the source impedance will vary in a power system subjected to random topology and environmental changes. This may, therefore, result in problems of stabilization for this feedback implementation and is a serious disadvantage of this technique.

With the template for the V_{gen}, a modulation technique to generate the firing pulses for the VSC switches is necessary. One of the most common methods is the Sinusoidal Pulse Width Modulation (SPWM) technique which uses a triangular carrier wave, as shown in Figure 2-18.

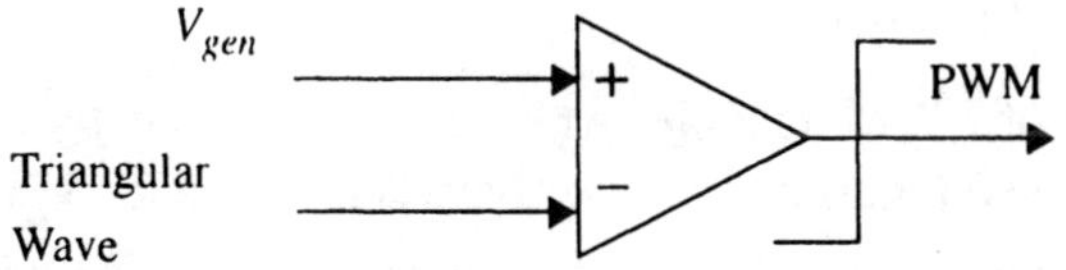

Figure 2-18: Sinusoidal PWM implementation

The two parameters that are of interest in a SPWM technique are the amplitude modulation index m_a and frequency modulation index m_f. Typical values of these parameters for high power converters are: $m_a = 0.8$ and $m_f = 15$. These values are chosen as a trade-off between reduced harmonic content in the output voltage and lower switching losses in the power electronic switches. An example of the PWM patterns obtained with these values are shown in the Figure 2-19 with either (a) bipolar or (b) unipolar voltage switching.

2.3.4.1 PWM with Bipolar Voltage Switching

In this PWM scheme, diagonally opposite switches from two legs of the converter are switched together as switch pairs 1 and 2, respectively. With this type of PWM switching, the output voltage waveform of the leg is identical to the output of the basic one-leg inverter, which is determined by comparison of a sinewave and a triangular wave. In the Figure 2-19a, it can be observed that the output voltage switches between $+V_d$ and $-V_d$ voltage levels. That is the reason why this type of switching is called PWM with bipolar voltage switching.

2.3.4.2 PWM with Unipolar Voltage Switching

In the PWM scheme with unipolar voltage switching, the switches in the two legs of the full-bridge inverter are not switched simultaneously, as in the previous PWM scheme. Here, the legs of the full-bridge inverter are controlled separately by comparing the triangular wave with a sine and cosine wave respectively. As shown in Figure 2-19b, the comparison of $V_{control}$ with the triangular waveform results in the logic signals to control the switches T1 and T3.

In this type of PWM scheme, when a switching occurs, the output voltage changes between zero and $+V_d$ or between zero and $-V_d$ voltage levels. For this reason, this type of PWM scheme is called PWM with a unipolar voltage switching, as opposed to the PWM with bipolar (between $+V_d$ and $-V_d$) voltage-switching scheme described earlier. This scheme has the advantage of "effectively" doubling the switching frequency as far as the output harmonics are concerned, compared to the bipolar voltage-switching scheme. Also, the voltage jumps in the output voltage at each switching are reduced to V_d, as compared to $2V_d$ in the previous scheme.

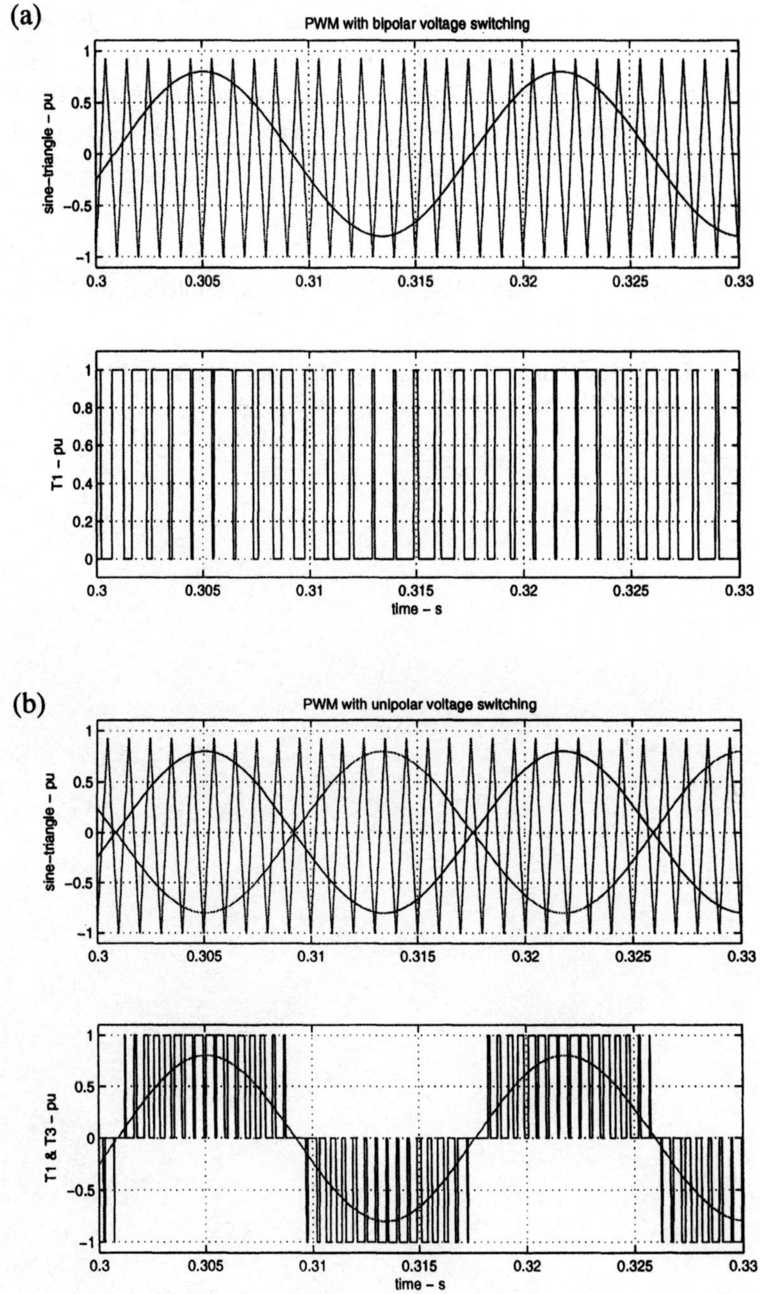

Figure 2-19: (a) Bipolar and (b) Unipolar switching techniques

2.4 CLOSING REMARKS

The traditional converter model which used to be based on the current source converter has now moved on towards a voltage source converter. This has opened the possibilities for FACTs type applications and the next decade will see increasing applications of this type.

2.5 REFERENCES

[1]. Direct Current Transmission, Vol. 1, E.W.Kimbark, Published by J.Wiley & Sons, ISBN 0-471-35550-X, 1971

[2]. Power Electronics Handbook, Edited by M.Rashid, Published by Academic Press, August 2001. ISBN 01258 16502, 1000 pages

Chapter 3

Synchronization Techniques for Power Converters

3.1 INTRODUCTION

The firing pulse generation unit of a static converter has a significant impact on the transient performance of the converter. For HVDC applications, a Voltage Controlled Oscillator (VCO) in conjunction with a Phase Locked Loop) is used to generate equi-distant firing pulses so that a satisfactory transient performance can be achieved even with relatively weak ac systems. In this chapter, the design and dynamic performance of two grid firing units, both based on Pulse Frequency Control (PFC), are assessed using EMTP.

Some of the earliest applications of converters in power systems were the HVDC transmission systems which assumed strong or infinite ac systems at the feeding source and the receiving ac system. However, as inter-connections to weaker ac systems became necessary in the mid-1960s, problems of harmonic instability were reported. One of the key elements in the control system of a static converter is the Grid Firing Unit (GFU). The application of Voltage Controlled Oscillators (VCOs) to HVDC converter firing control was first described in [1]. This method derived equi-distant firing pulses for the converter valves and, to a large extent, decoupled the impact of any commutation ac voltage distortion on the valve triggering [2]. This made it feasible for the successful operation of converters with relatively weak ac systems having Short Circuit Ratios (SCRs) of less than 3. Increasingly, as economic pressures force utilities to consider still weaker ac systems (with SCRs of less than 2), problems of harmonic instability are being experienced.

One common type of GFU, hereafter referred to as the Conventional type, is based on a VCO in conjunction with a PLL [5]. In the circuit, the synchronizing voltage V_{sync} is compared with the commutation voltage V_{com} from the ac system bus. The error between these two signals is then fed to a VCO to alter the frequency and phase angle of the synchronizing voltage such that this error is reduced to zero.

Another type of GFU, referred to as the *Transvektor* type [6] or DQO type, has a DQO transformation stage in the circuit. This DQO-type has been used in motor drives applications for many years. Its first application to an HVDC system was made at Chateauguay, Quebec in the middle 1980's. One advantage claimed of the DQO-type GFU was its superior immunity to disturbances and harmonic distortion. However, a detailed comparison of these two types has been previously reported [9] and no major differences were observed.

In this chapter, the design and performance of these two types of GFUs under steady state and dynamic conditions is presented. The operational principles of these two circuits are briefly reviewed and the procedure of designing the circuits are described. The circuits are then tested under fault and harmonic distortion conditions. An HVDC rectifier system based on the CIGRE benchmark system [7] is studied using EMTP and results are provided for comparison.

3.2 REVIEW OF GFUs

The primary objective of a GFU is to provide firing pulses to the converter valves in the correct phasor relationship to the relevant fundamental component of the commutation voltage, V_{com}. In the case of a strong ac system, obtaining a stable pollution-free V_{com} signal is usually not a problem. However, in a weak ac system, it is difficult to obtain a stable pollution-free V_{com} signal. A suitable GFU, therefore, has the task of deriving a clean pollution-free synchronizing voltage signal V_{sync} which must be in an exact phasor relationship with the fundamental component of the commutation voltage. During transients and disturbances, the V_{sync} signal may temporarily lose synchronism with the V_{com} signal. The dynamic characteristics of the GFU must be capable of restoring synchronism rapidly, and in a stable manner, with the commutation voltage as soon as the transient is over.

There are two types of GFUs that have been widely used; one based on Individual Phase Control (IPC) and the other on Equi-Distant Pulse Control (EPC).

3.2.1 Individual Phase Control (IPC) Unit

In this type of GFU (now obsolete), the firing pulses are directly derived from the zero crossover points of the commutation voltage. Consequently, the firing pulses are vulnerable to harmonic pollution on the waveform. Early attempts to use filtering techniques to alleviate some of these problems were not successful for operation with weak ac systems due to the introduction of phase shifts. Developments in tracking band-pass filters [8] which derive the fundamental frequency component of the commutation voltage with no phase shift may be useful in operation with weak ac systems. However, the main disadvantage of IPC systems, which eventually led to their demise, was the generation of non-characteristic harmonics which caused harmonic instability problems.

3.2.2 Equi-Distant Pulse Control (EPC) Unit

EPC systems generate only characteristic harmonics during steady state operation. Two GFUs of this type have been described in the literature:

3.2.2.1 Pulse Frequency Control (PFC) Type

To decouple the direct dependence of the pulse firing from the zero crossover points of the commutation voltage, a VCO followed by a ring counter is used [1,3]. The characteristic feature of this method is that a dc input control signal to the VCO results in a change in the frequency of the VCO. For this reason, this type of GFU is referred to as of the PFC type [4].

A free running VCO generates a train of short pulses at a pulse repetition frequency directly proportional to the dc control voltage. For example, if the control voltage is adjusted such that the oscillator frequency is at 6 (or 12) times the ac supply frequency, then the pulses will be exactly at 60° (or 30°) intervals (and hence the term equi-distant firing pulses). A ring counter is used to separate the pulse train into 6 (or 12) sets of pulses for a 6 (or 12) - pulse converter.

An indirect method is used to synchronize the VCO output frequency to the ac supply frequency. An error signal is derived from either the converter dc current or extinction angle controller as a feedback signal. When the error signal is zero, the GFU is in steady state (free running or at a centre frequency) and the VCO output frequency will be at supply frequency. When there is an error, the VCO will either speed up or slow down to correct for the error. Both [1] and [3] use the dc current controllers for synchronizing the VCO when the converter is a rectifier; however, when the converter is an inverter, the VCO synchronizing is based on an extinction angle controller, although the method used by [3] is based on a predictive estimation of extinction angle.

Due to the integral characteristic of the VCO, it is not possible to modulate the firing pulses on an individual basis; for this reason, an asymmetric firing unit is used in [3] to optimize dc power flow during unbalanced ac system faults.

3.2.2.2 Pulse Phase Control (PPC) Type

In a GFU of this type [4], the dc control voltage resulted in a change to the phase of the VCO output rather than its frequency. The transfer function of this type of unit is therefore proportional rather than integral. To ensure the synchronism of the VCO output frequency with the ac supply frequency, a slower acting frequency error feedback loop is used. This type of GFU does not permit the modulation of firing pulses on an individual basis either.

3.3 GFUs - DESIGN AND ANALYSIS

3.3.1 Conventional GFU

The block diagram of a conventional GFU [9] is shown in Figure 3-1.

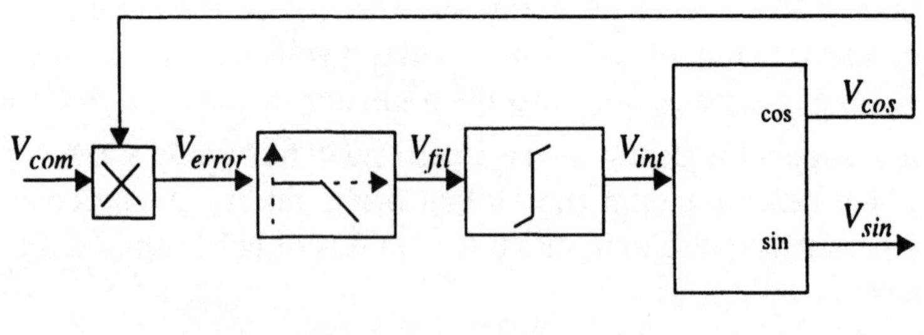

Figure 3-1: Block diagram of conventional GFU

In this circuit, the commutation voltage, assumed to be $V_{com} = 1sin(\omega_1 t + \theta_1)$, is multiplied by a feedback signal, $V_{\cos} = 1cos(\omega_2 t + \theta_2)$. The output voltage V_{error} is obtained according to eq.(3-1).

$$V_{error} = 1\sin(\omega_1 t + \theta_1)\cdot 1\cos(\omega_2 t + \theta_2)$$

$$V_{error} = 0.5\sin([(\omega_1 - \omega_2)t + (\theta_1 - \theta_2)] + 0.5\sin[(\omega_1 + \omega_2)t + (\theta_1 + \theta_2)])$$

(3-1)

The first term of eq.(3-1) represents the error between the synchronizing voltage and the commutation voltage due to the frequency and phase difference. Under steady state, the synchronizing voltage will be locked to the commutation voltage. In this case, $\omega_1 = \omega_2$ and $\theta_1 = \theta_2$, and the first term of eq.(3-1) is zero. The second term is an unwanted ac component which has a frequency of $2\omega_1$ under steady state. In order to extract the dc error signal and filter out the unwanted ac component, a low-pass filter having the transfer function $\omega_c/(s+\omega_c)$ is used. The output is passed onto an integrator with a transfer function of $1/sT_i$. The integrator output, V_{int}, is used to modulate the phase and frequency of a free-running Sine-Cosine Oscillator to generate the output signal V_{sync}. Under steady state conditions, the feedback signal V_{sync} will be in phase and at the same frequency as the commutation voltage, V_{com}. Thus V_{sync} can be used as a stable pollution-free signal to derive the zero-crossover points to provide the timing reference points for the GFU.

Figure 3-2 shows the waveforms of the conventional GFU shown in Figure 3-1. It can be seen that the error signal, V_{error}, contains a dominant second harmonic ac component which is removed by a low-pass filter to reduce its impact on the overall system operation. The bottom superimposed signals in Figure 3-2 show the commutation voltage V_{com} and the synchronizing voltage V_{sync}. Note that, in order to see clearly the phase difference of the two signals, the magnitude of V_{sync} has been deliberately increased by 20%.

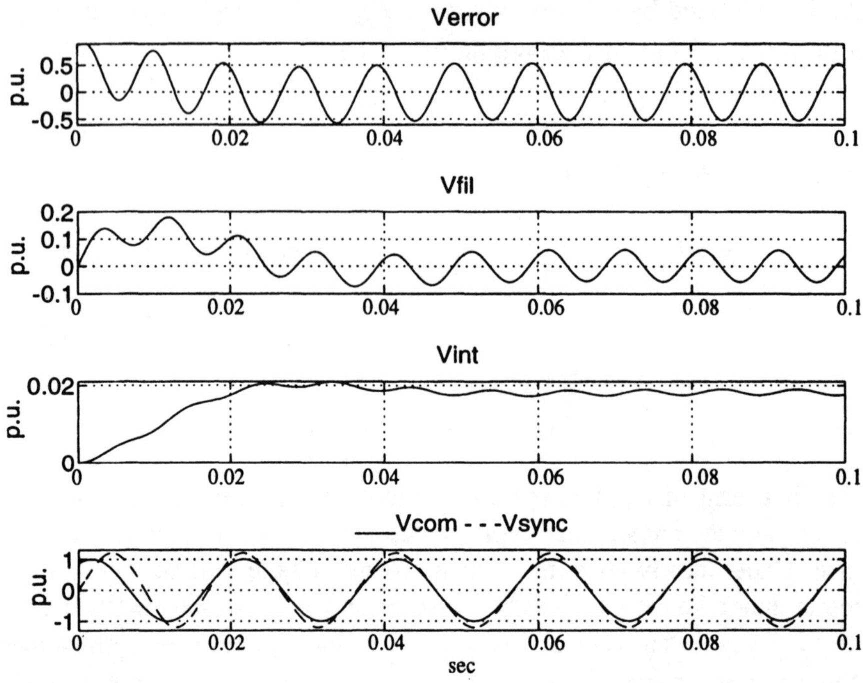

Figure 3-2: Waveforms of conventional GFU

There are two parameters that need to be designed in the circuit of Figure 3-1. One is the cut-off frequency of the low-pass filter, and the other is the integrator time constant. The design objective is to achieve synchronization between V_{sync} and V_{com} in the shortest time possible. One common design approach is to study the small signal model of the circuit and design the parameters in the frequency domain. Studies show [9] that the small signal model of the circuit in Figure 3-1 can be represented as shown in Figure 3-3, where *H(s)* is the low-pass filter transfer function and variables in hats represent small signal quantities. The loop transfer function is given in eq.(3-2).

$$T_{l_2} = \left(\frac{\omega_c}{s+\omega_c}\right)\left(\frac{1}{sT_{i_i}}\right) \tag{3-2}$$

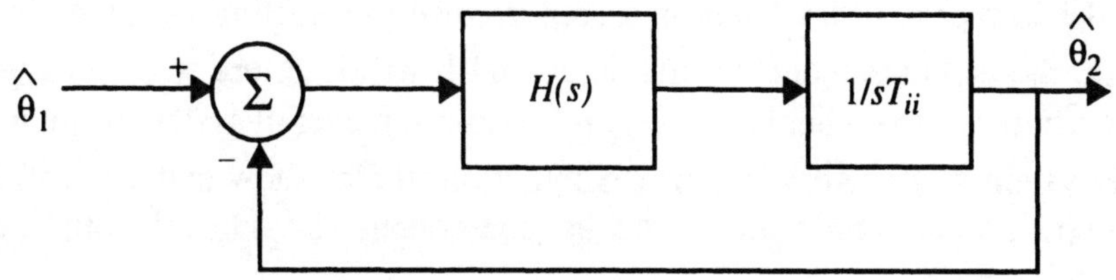

Figure 3-3: Small signal model of conventional GFU

Figure 3-4 gives the Bode plot of the loop transfer function. In this figure, the solid line represents the response of the loop transfer function $T_{l2}(s)$, and the dotted lines are for the low-pass filter function *H(s)* and the integrator function $1/sT_{ii}$. This figure shows that, in order to achieve the optimum phase margin of around 60°, the integrator time constant should be selected such that the value of $1/T_i$ is smaller than ω_c, the cut-off frequency of the low-pass filter.

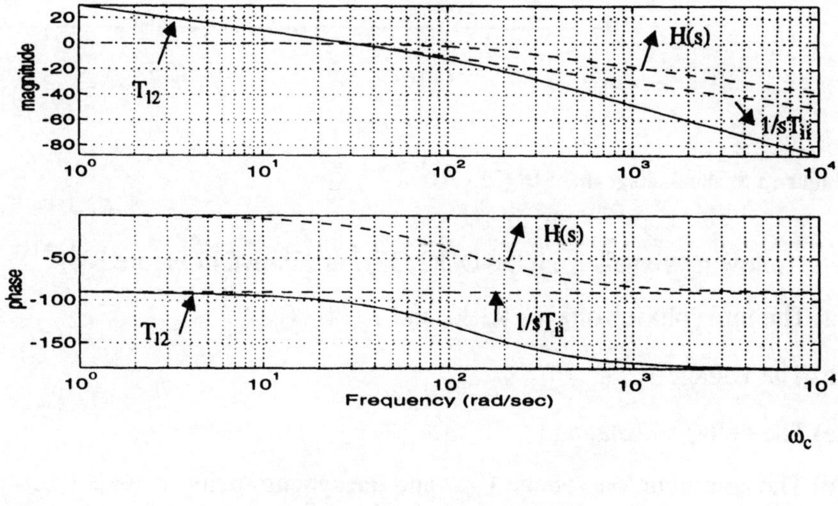

Figure 3-4: Bode plot for the loop transfer function of eq.(3-2)

Figure 3-4 also shows that the loop response speed, represented by the gain-crossover frequency, is to a large extent, limited by the cut-off frequency ω_c of the low-pass filter. There is a compromise in selecting ω_c. If ω_c is too high, the ac component remains large and it will interfere with the system operation. On the other hand, if ω_c is too low, the overall system response of the system will be very sluggish. Simulation studies show that a cut-off frequency around one fifth of the ac component ($2\omega_c$) gives satisfactory results.

3.3.2 DQO GFU

The block diagram of the DQO GFU is shown in Figure 3-5.

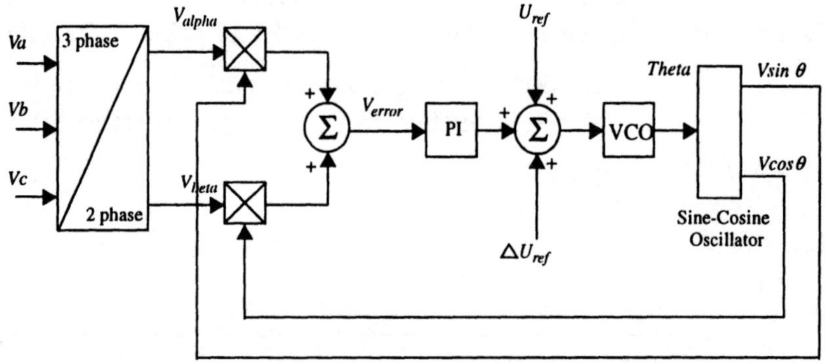

Figure 3-5: Block diagram of DQO GFU

The following signals from the DQO GFU are shown in Figure 3-6:

(a) The three phase voltages V_a, V_b and V_c,

(b) The voltage V_{error},

(c) The voltage *Theta*, and

(d) The commutation voltage V_{com} and the synchronizing voltage V_{sync}.

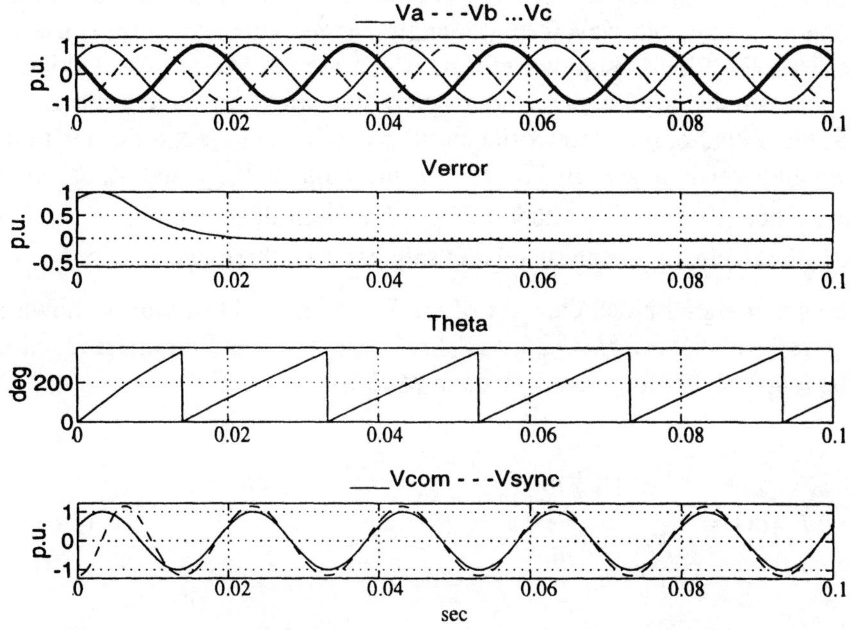

Figure 3-6: Waveforms of DQO GFU

The three phase commutation voltages V_a, V_b and V_c are transformed into DQO axis voltages V_{alpha} and V_{beta} using eq.(3-3) and eq.(3-4) respectively.

$$V_{alpha} = \left(\frac{2}{3}\right)V_a - \left(\frac{1}{3}\right)V_b - \left(\frac{1}{3}\right)V_c \tag{3-3}$$

$$V_{beta} = \left(\frac{1}{\sqrt{3}}\right)(V_b - V_c) \tag{3-4}$$

$$V_{error} = (V_{alpha} V\sin\theta) + V_{beta} V\cos\theta) \tag{3-5}$$

An error signal, V_{error}, derived using eq.(3-5), is fed through a PI controller to generate a reference value for the VCO. This reference value can be modulated by a signal ΔU_{ref}, and it has a fixed voltage bias U_{ref} which sets the center frequency of the VCO. The output of the VCO is a signal proportional to a sawtooth waveform (an angle *Theta*). This waveform is used to generate the Sine-Cosine waveforms which are fed back to the multipliers to generate the error signal. Under steady state, this error is reduced to zero and the output of the Sine-Cosine oscillator will be in synchronism with the commutation voltages. In Figure 3-6, the outputs V_{sync} and V_a are compared; note that the magnitude of V_{sync} has been deliberately increased by 20% to see clearly the phase relationship between these two signals.

The small signal block diagram of the DQO grid control unit is shown in Figure 3-7, where $G(s) = K_{pi}(1+sT_{pi})/sT_{pi}$ represents the PI transfer function. The loop transfer function is given in eq.(3-6).

$$T_{l2}(s) = K_{pi}\left(\frac{1+sT_{pi}}{sT_{pi}}\right)\left(\frac{1}{sT_i}\right) \tag{3-6}$$

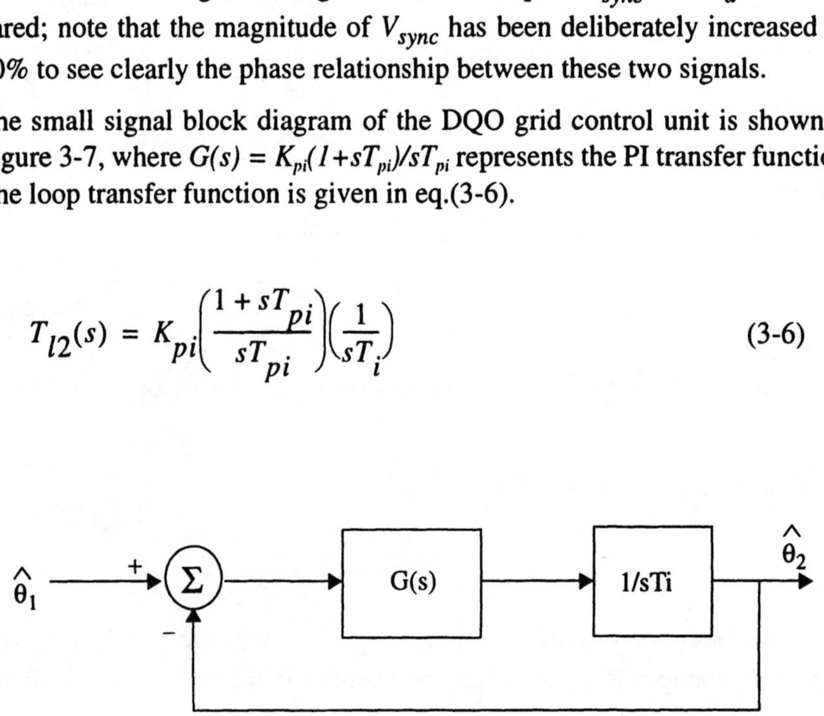

Figure 3-7: Small signal block diagram of DQO grid unit

The Bode plot of $T_{l2}(s)$ is shown in Figure 3-8. The solid lines are for the loop transfer function $T_{l2}(s)$ and the dotted lines are for the PI controller and the integrator transfer functions. From Figure 3-8, it can be concluded that, in order to achieve the optimum phase margin of around 60°, the value of $1/T_i' = K_{pi} / T_i$ should be larger than $1/T_{pi}$. Also, since the phase lag approaches 90° as the frequency increases, theoretically the gain-crossover frequency can be chosen as high as one wishes. Practically, this frequency will be limited in a realistic system. One of the limiting factors is the existence of the low order harmonic component in the V_{error} signal when the three-phase ac source contains harmonics. For example, with a third harmonic injection at the ac bus, the V_{error} signal will contain a second harmonic ac component. Under such operating conditions, as the gain crossover frequency increases, the error between the synchronizing signal (V_{sync}) and the fundamental component of the commutation voltage (V_{com}) increases as well. Our studies show that the gain crossover frequency of around 40 Hz provides a good compromise between a fast response and a small synchronizing error.

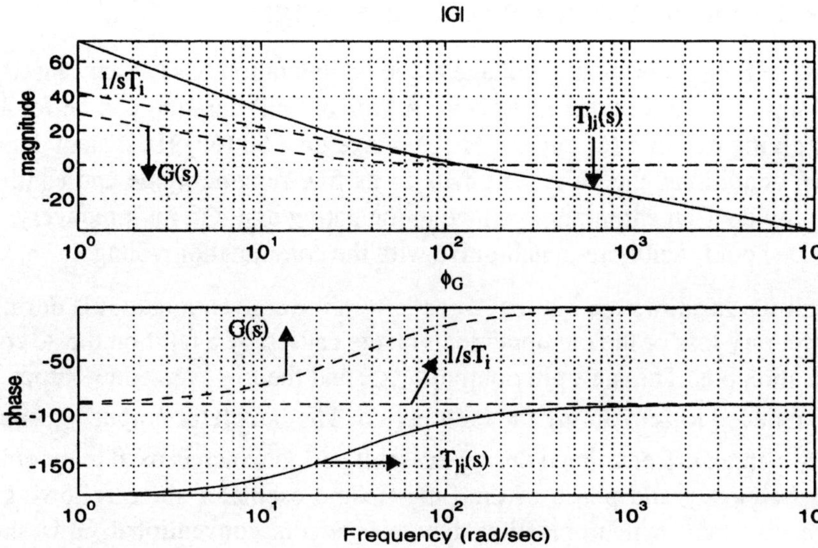

Figure 3-8: Bode plot for the loop transfer function of eq.(3-6)

3.3.3 Comparison

The major difference between the operational behaviors of the conventional and DQO GFUs is the presence of the ac harmonic component in the V_{error} signal under normal operating conditions. In the conventional GFU, a large second harmonic ac component exists and a low-pass filter is required in the loop. While in the DQO GFU, there is no such ac component in the V_{error} signal. The absence of the low-pass filter allows the DQO circuit to achieve a relatively faster dynamic response than its conventional counterpart. Experience also shows that the optimization of the low-pass filter requires additional effort when compared to the DQO circuit.

3.4 TESTS ON GFUs

The behavior of the GFUs during system disturbances causing a loss of commutation voltage and in the presence of harmonics is discussed in this section.

3.4.1 Loss of Synchronization Voltage

Loss of the synchronizing voltage at the commutation bus can be caused by a temporary fault on the ac system. A typical fault duration can be for 100 ms giving rise to 5 (6) cycle loss of voltage on a 50 (60) Hz system. Under such conditions, the GFU falls back to its free-running mode and continue to provide a synchronizing voltage to the gating unit. On fault recovery, the GFU should rapidly re-synchronize with the commutation voltage.

Figure 3-9a shows the internal signals from the conventional GFU during a temporary loss of the commutation voltage caused by a fault on the ac commutation bus. The multiplier output V_{error} and the low-pass filter output V_{fil} are reduced to zero during the fault period. The Integrator output V_{int} shows only a small offset voltage during the fault period which is used to modulate the frequency and phase of the Sine-Cosine oscillator stage following it. The post-fault synchronization dynamics of the conventional GFU show that the output voltage V_{sync} is able to synchronize with the commutation voltage V_{com} within 1 cycle (20 ms at 50 Hz). The waveform of V_{int} also shows that the control loop is slightly under damped and requires a settling time of about 3 cycles.

Figure 3-9b shows the internal signals from the DQO GFU during a temporary loss of the commutation voltage caused by a three phase fault on the ac commutation bus. During the fault, the three phase commutation voltages V_a, V_b and V_c are reduced to zero causing the V_{error} input to the PI controller to drop to zero. This results in the output of the sawtooth waveform, *Theta*, to be at the centre frequency (50 Hz in this case). After the fault, the error is reduced to zero within 1 cycle (20 ms).

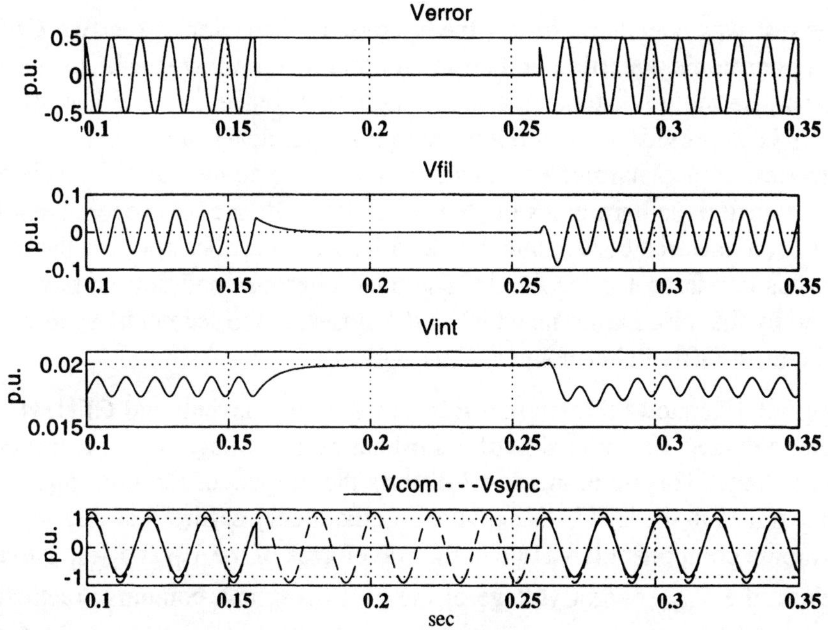

Figure 3-9a: Loss of synchronization voltage for conventional unit

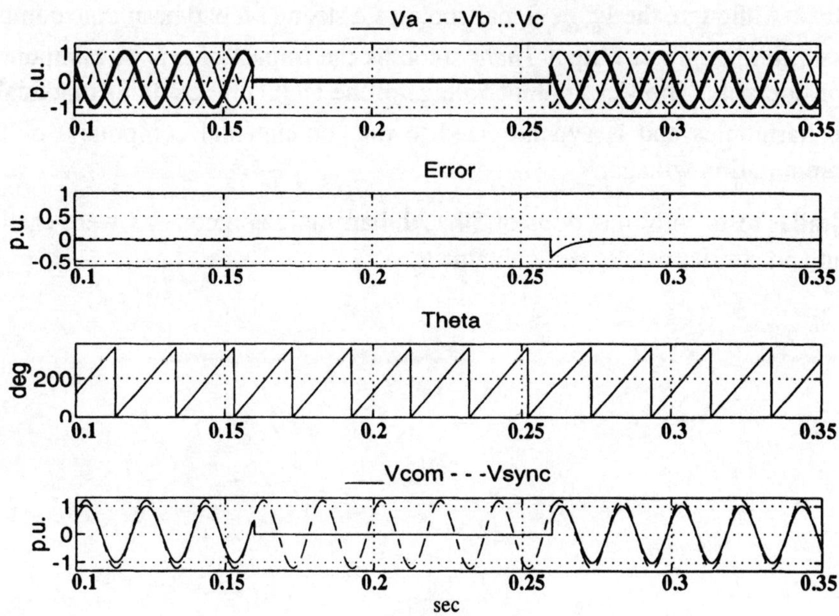

Figure 3-9b: Loss of synchronization voltages for DQO unit

3.4.2 Harmonic Distortion Test

This test was conducted to test the operational behavior of the two GFUs under harmonic distortion at their input commutation voltage. In a weak ac system, harmonic distortion is a common occurrence, and the role of the GFUs is to provide a clean output with minimal delay for synchronization purposes. Typical harmonics that are present in the commutation voltage are the characteristic harmonics i.e. 5th, 7th, 11th, 13th etc. However, even the lowest characteristic harmonic, i.e. the 5th harmonic, does not pose the most onerous condition for the GFUs. The most onerous condition is, however, posed by the third harmonic which is the closest to the second harmonic and is the most difficult harmonic to filter out.

Figure 3-10a shows the internal signals from the conventional GFU with a rather elevated 30% injection of a third harmonic voltage on the commutation voltage. This harmonic level distorts the outputs of the multiplier and other stages in the GFU. As explained in the theory earlier, a strong second harmonic component is visible in the waveforms of the V_{fil} and V_{int}. Nevertheless, the V_{sync} output voltage of the gird firing unit contains practically no harmonics and is synchronized to the fundamental component of the commutation voltage.

Similarly, Figure 3-10b shows the corresponding signals with the DQO GFU. Although, the V_{error} signal contains a strong second harmonic component, the integrator output *Theta* smooths the impact of this ac component considerably. The V_{sync} output voltage of the DQO unit contains practically no harmonics and is synchronized to the fundamental component of the commutation voltage.

Similar tests with injections of 5th, 7th harmonic components were carried out and similar results were obtained.

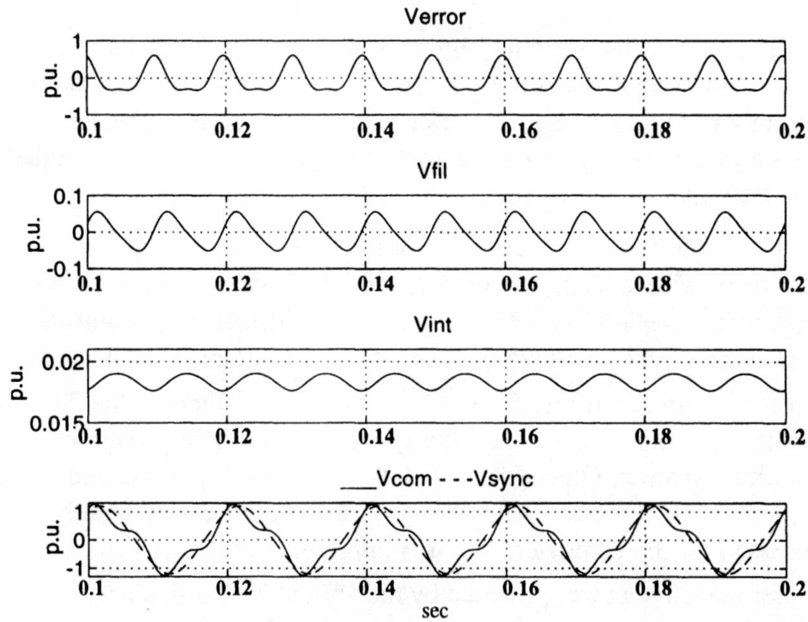

Figure 3-10a: Harmonic distortion test with conventional unit

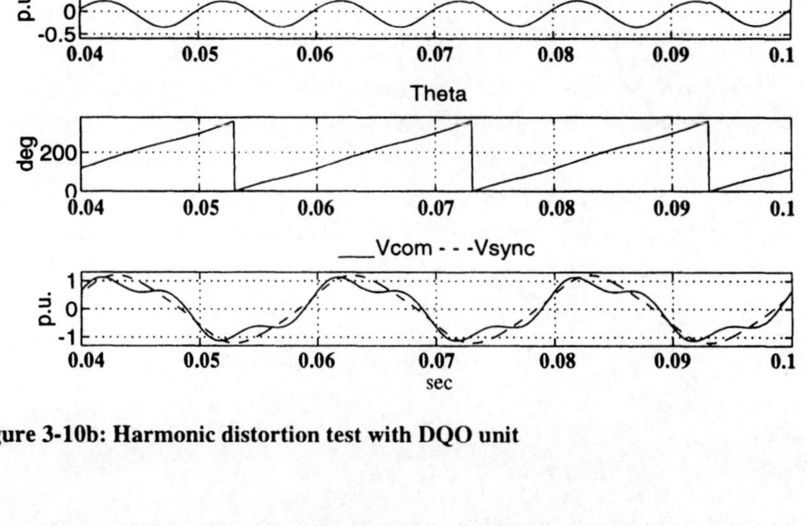

Figure 3-10b: Harmonic distortion test with DQO unit

3.5 EMTP SIMULATION OF A TEST SYSTEM

An HVDC rectifier system (Figure 3-11), loosely based on the CIGRE Benchmark system [9], is used as the test system. A 6-pulse model is used here only to minimize the simulation time. However, the same design principles and the operational characteristics for the GFUs can be extended to a 12-pulse unit.

The fixed frequency ac source has an impedance formed by the 2R-L network to provide a short circuit ratio of 2.5 at the commutation bus. This constitutes a weak ac system and provides a difficult synchronization task for the GFUs. Since a 6-pulse version of the converter system is modelled, it is necessary to add tuned 5th and 7th harmonic ac filters at the 345 kV ac commutation bus to cope with the generation of characteristic harmonics from the converter. Other filters include a damped high-pass and a double tuned filter. A capacitor bank also provides the necessary reactive power consumed by the converter; this was appropriately reduced to account for the extra reactive power provided by the 5th and 7th harmonic ac filters.

The converter transformer is a grounded star-star transformer with its leakage impedance split equally between its primary and secondary windings. The saturation characteristic from the transformer has a knee level at 1.2 pu.

The dc system is composed simply of a smoothing reactor with its inherent resistance and a resistive load. A shorting switch is included for applying a dc line fault test case.

Complete data for the system modelled with EMTP is presented in the figure.

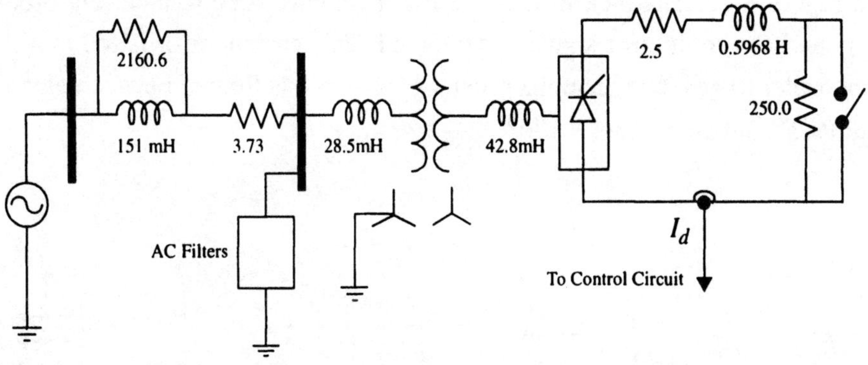

AC FILTER DATA:

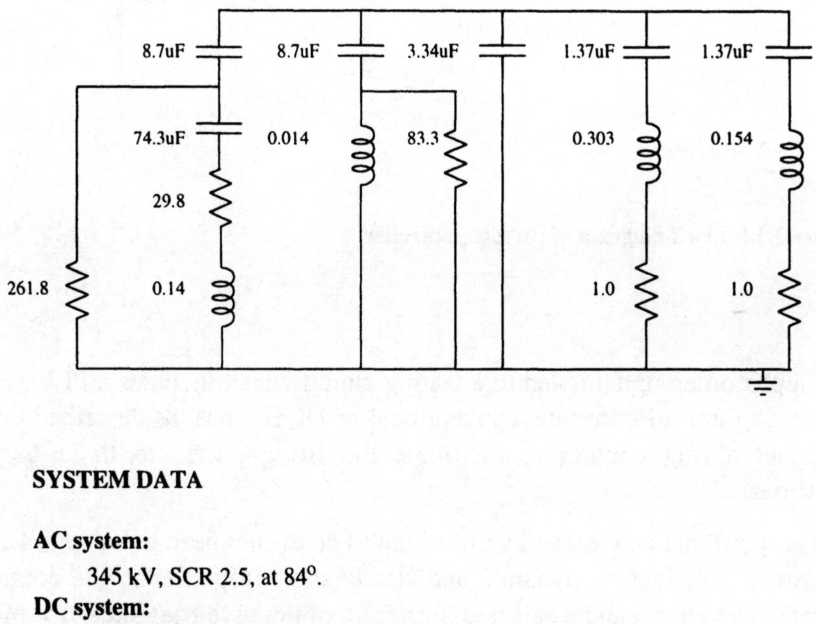

SYSTEM DATA

AC system:

345 kV, SCR 2.5, at 84°.

DC system:

$P_d = 1000$ MW, $V_d = 500$ kV, $I_d = 2000$ A, $R_L = 250\ \Omega$.

Converter Transformer:

Rating 2x598 MVA, impedance 18%, turns ratio 0.612

Saturation at 1.2 pu knee level

Smoothing reactor:

$L_d = 596.8$ mH, $R_d = 2.5\ \Omega$

Figure 3-11: The rectifier system used for studies based on the CIGRE Benchmark model

The block diagram of the current controller used with this system is shown in Figure 3-12. The measured dc current I_d is compared to a current order I_{ref}, and a current error signal is generated. This current error is fed to a PI controller to generate an alpha order signal which is limited between alpha-min = 5° and alpha-max = 170°.

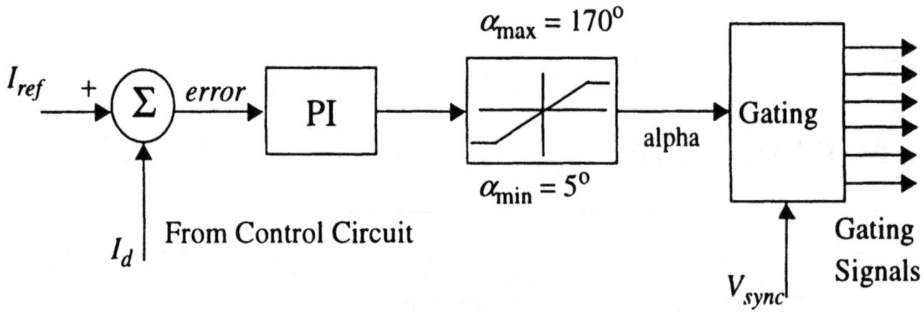

Figure 3-12: Block diagram of current controller

The alpha order signal is fed to a Gating circuit which includes a PLL synchronizing unit of either the conventional or DQO types, as described earlier, and a ring counter to distribute the firing pulses to the 6-pulse converter.

The proportional and integral gains of the PI controller need to be optimized to provide satisfactory dynamic and steady state properties to the control system. The gains can be selected as part of an iterative trial-and-error process with the tests that will be discussed next. Optimization of these gains is a non-trivial exercise.

3.5.1 Start-up Of System Model

Figures 3-13a and 3-13b show the start-up of the rectifier system with the conventional and DQO GFUs, respectively. No special initialization procedures are utilized for these cases in order to study the start-up process and observe the stability characteristics of the overall system. Particular attention should be paid to the very first cycle of the synchronizing and commutation voltages. The mere realization that the system will actually initialize and start-up is achievable is often a very good indication for the simulation.

The signals shown are the dc current I_d, alpha order and the superimposed waveforms of the synchronizing voltage ($1.2*V_{sync}$) and the commutation voltage V_{com}. Note that the synchronizing voltage has been deliberately increased by a factor of 1.2 pu to illustrate and compare more clearly the time related behavior of the two sinewaves V_{sync} and V_{com}.

In both cases, the start-up of the dc system was achieved rapidly and the outputs of the GFUs are able to synchronize with the commutation voltage within 1 cycle. It is also important to adjust the controller gains of the GFUs to achieve a dynamic performance capable of synchronization within 1 cycle. The dynamic performance should, in principle, be much faster than the response of the main current (or gamma) controller of the converter; typically, these response times are in the 3-5 cycle region. A more detailed examination shows that the DQO GFU is marginally faster than the conventional unit.

From the perspective of the main current controller, in both cases, the alpha order picks up from its alpha-minimum value of about 5 degrees and achieves its final steady state value of about 22 degrees; in reality, this is higher than normal as the final nominal value is typically around 17 degrees. The dc system achieves full load transmission in about 6 cycles.

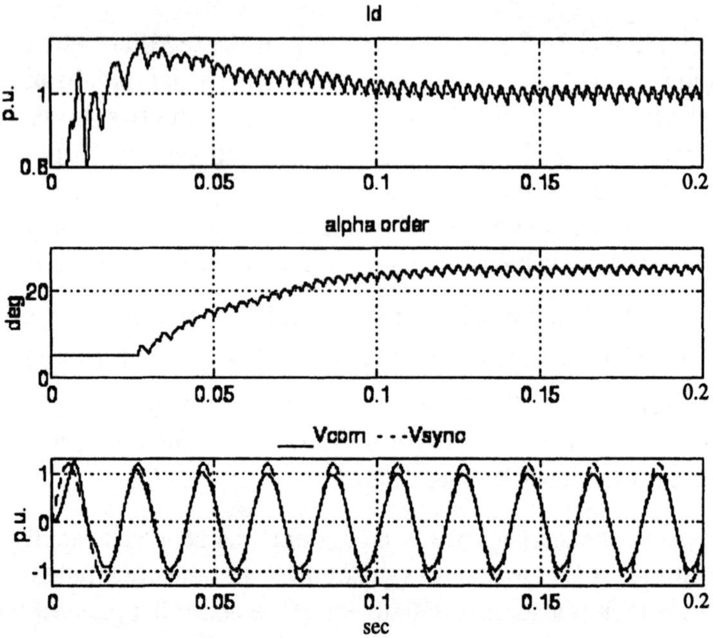

Figure 3-13a: Initialization with conventional GFU

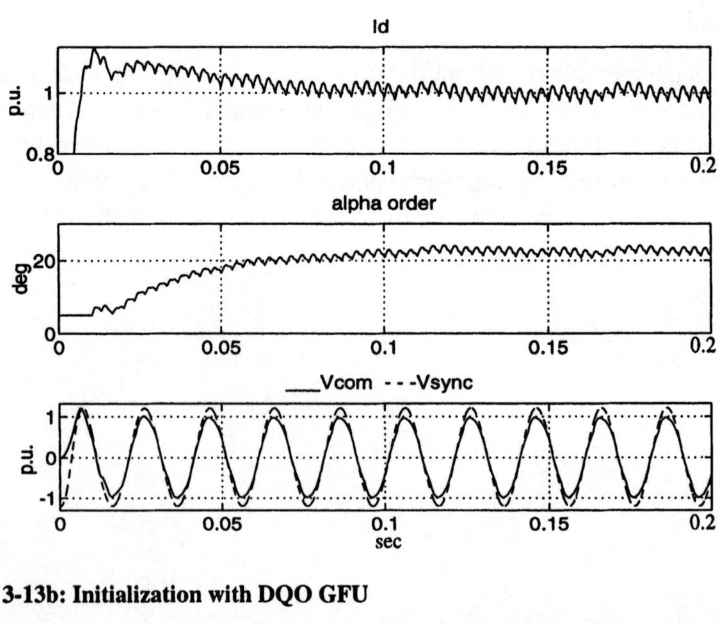

Figure 3-13b: Initialization with DQO GFU

3.5.2 10% Step Change In Current Order

Small step changes in the reference values are often used to test the stability and dynamic responses of the main converter controllers. The dynamic responses of the GFUs are expected to be an order of magnitude faster than the main controllers so as not to interfere with the dynamic behavior of the main controllers. However, in reality the GFUs are only about 3 to 5 times faster, and may interact with the behavior of the main controller, especially when operating with weak ac systems. The relative strength (or short circuit ratio) of the ac system has an impact on the speed of response and stability of the two controllers. Increasingly, utilities are veering towards weaker ac systems, and the improved performance expectations of the GFUs and the main controllers are increasing.

Figures 3-14a and 3-14b show the case of a 10% step change in current order for the rectifier system with the conventional and DQO GFUs respectively. The signals shown are the dc current I_d, alpha order and the superimposed values of the synchronizing voltage ($1.2*V_{sync}$) and the commutation voltage V_{com}. The step reduction is applied at 0.25 seconds for a duration of 200 ms.

For the conventional unit, the step change is effected in 30 ms, and the response is well controlled and stable.

For the DQO unit, the step change is somewhat slower and takes about 50 ms. Also, unfortunately, a persistent fundamental frequency (50 Hz) component is visible in the dc current for the DQO unit. It is unclear why this harmonic component persists in the simulation, and further tests need to be conducted to investigate this. It is possible that better results may be obtained with further optimization of the controller parameters.

However, what is relevant here is that both GFUs are able to maintain synchronism with the commutation voltage without any noticeable delays.

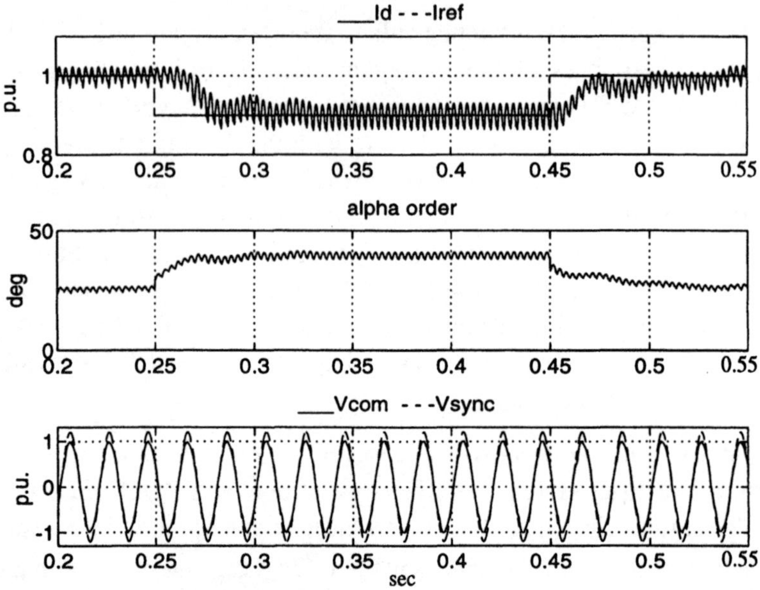

Figure 3-14a: 10% current step change with conventional GFU

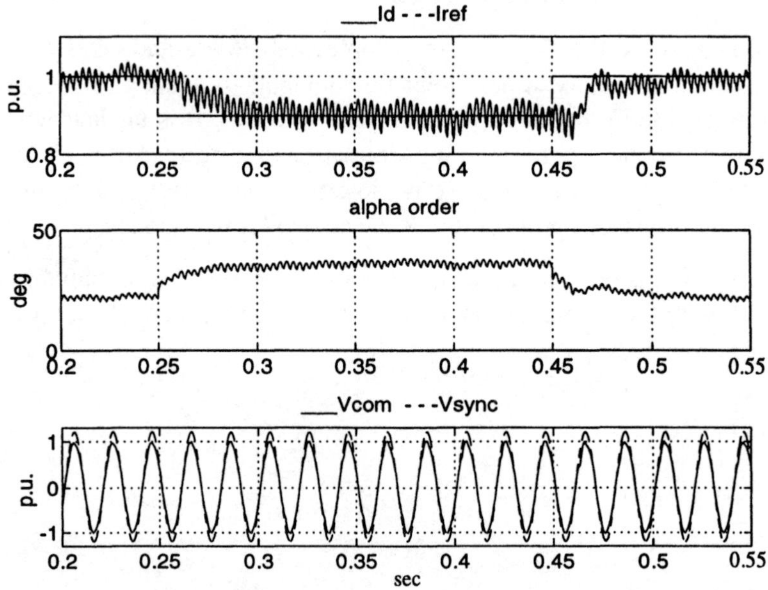

Figure 3-14b: 10% current step change with DQO GFU

3.5.3 Single Phase Fault

A remote single-phase, unbalanced ac fault is the most common type of perturbation on a power system. Typical durations of such faults are in the 3-10 cycles range. Usually, the dc controllers are designed to simply ride through these types of disturbances with a marginal reduction in the power flow. However, one characteristic of such a fault is worthy of consideration from the point of view of the GFU behavior. During the unbalanced fault, the modulation effect of the non-linear converter generates a strong second harmonic component which is often not eliminated by the low pass filtering characteristic of the GFUs. This test is, therefore, a good verification of the capabilities of the GFUs and their ability to operate under such circumstances.

Figures 3-15a and 3-15b show the case of a single phase fault for the rectifier system with the conventional and DQO GFUs respectively. The signals shown are the dc current I_d, alpha order and the superimposed waveforms of the synchronizing voltage ($1.2*V_{sync}$) and the commutation voltage V_{com}. The fault period is from 0.25 to 0.45 seconds i.e. a fault duration of 200 ms or 10 cycles at 50 Hz.

In both cases the responses of the main dc controllers are well controlled and stable, although the case with the DQO unit is marginally faster. The presence of the large second harmonic component in the dc current during the fault period demonstrates the inability of the main current controllers to filter out completely the impact of the second harmonic. However, this does not effect the operation of the grid firing units noticeably and they maintain synchronism throughout the fault period. It is difficult to ascertain which GFU is better under these conditions.

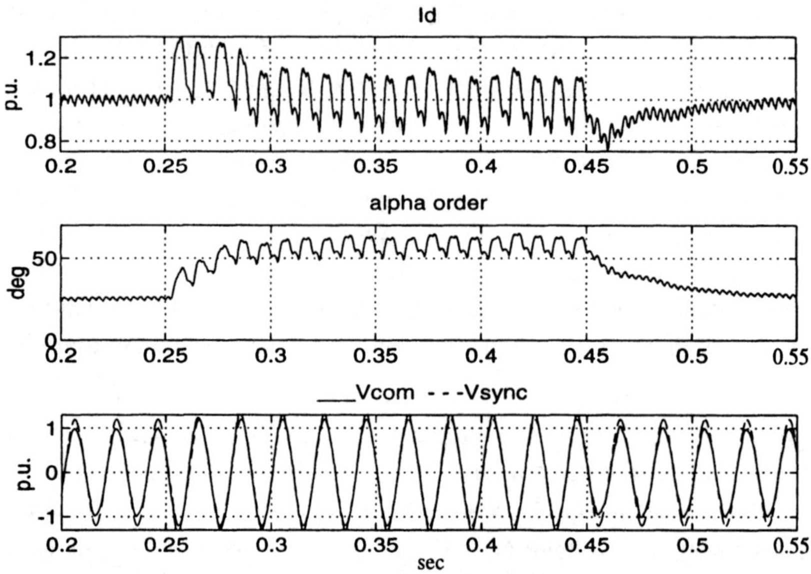

Figure 3-15a: Single phase fault with conventional GFU

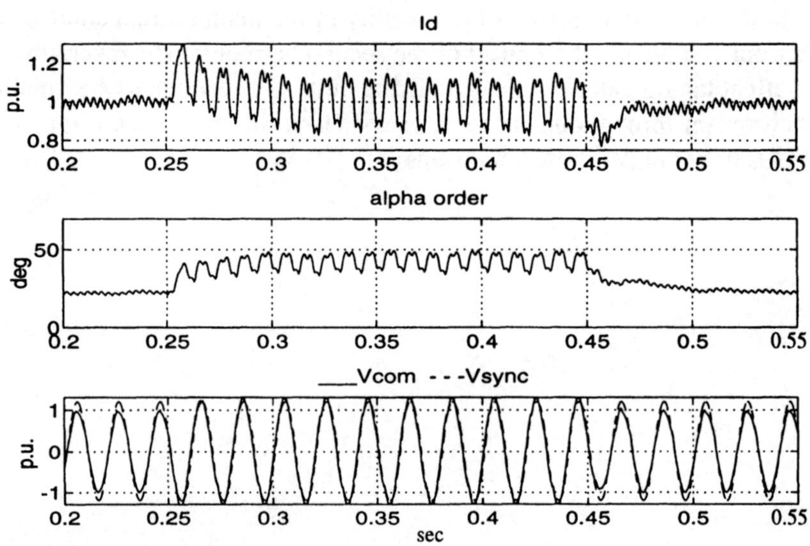

Figure 3-15b: Single phase fault with DQO GFU

3.5.4 DC Line Fault

A dc line fault in a HVDC system results in a total load rejection from the point of view of the rectifier ac system. This leads to a large rise in the commutation bus voltage due to the availability of surplus reactive power from the capacitor and filter banks at this bus. This voltage rise can often exceed 1.8 pu, and can cause saturation of the converter transformer which, typically, has a knee level at 1.2 pu. The transformer saturation results in harmonic generation which can impact on the performance of the GFUs and the main controllers.

Figures 3-16a and 3-16b show the case of a dc line fault for the rectifier system with the conventional and DQO GFUs respectively. The signals shown are the dc current I_d, alpha order and the superimposed waveforms of the synchronizing voltage ($1.2*V_{sync}$) and the commutation voltage V_{com}. Note that in these two cases, no protective Voltage Dependent Current Limits (VDCL) are utilized to reduce the current order - typical in a practical plant - after the fault inception since only the control response of the GFUs is under investigation here. In these cases, therefore, the current order was maintained at 1.0 pu during the post-fault period. Hence in the results shown, the dc current is regulated back to 1 pu in the post fault period within 2 cycles.

In both cases, the GFUs are able to maintain synchronism within 1-2 cycles of the fault inception. The loss of the dc load causes the commutation voltage to rise and generate harmonics due to transformer saturation, but this does not cause persistent problems for the GFUs to generate a clean synchronizing voltage. In the case of weak ac systems, this commutation voltage rise and the subsequent generation of harmonics is particularly important for the main current controllers.

From a point of view of the dc controllers, they were able to limit the peak fault dc current to about 2.5 pu while the alpha order increased rapidly to about 145 degrees to limit the peak fault current.

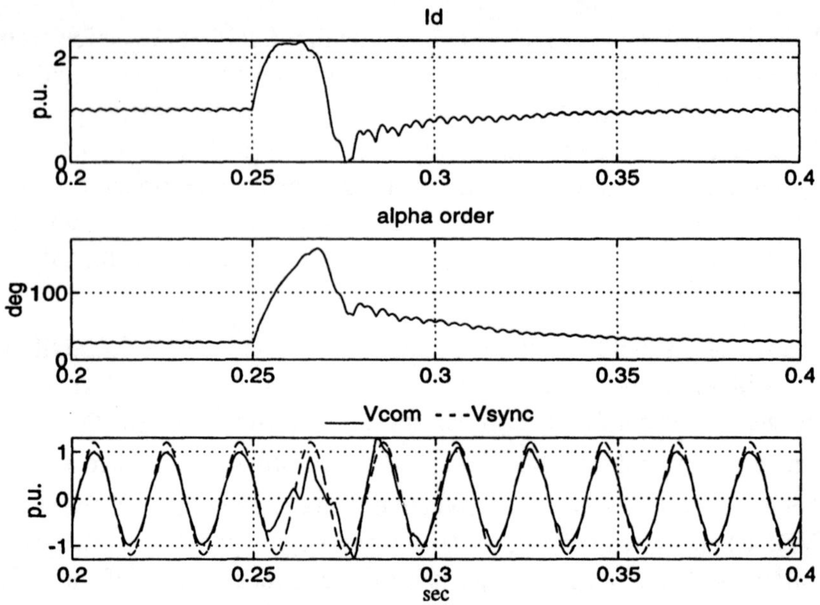

Figure 3-16a: DC line fault with conventional GFU

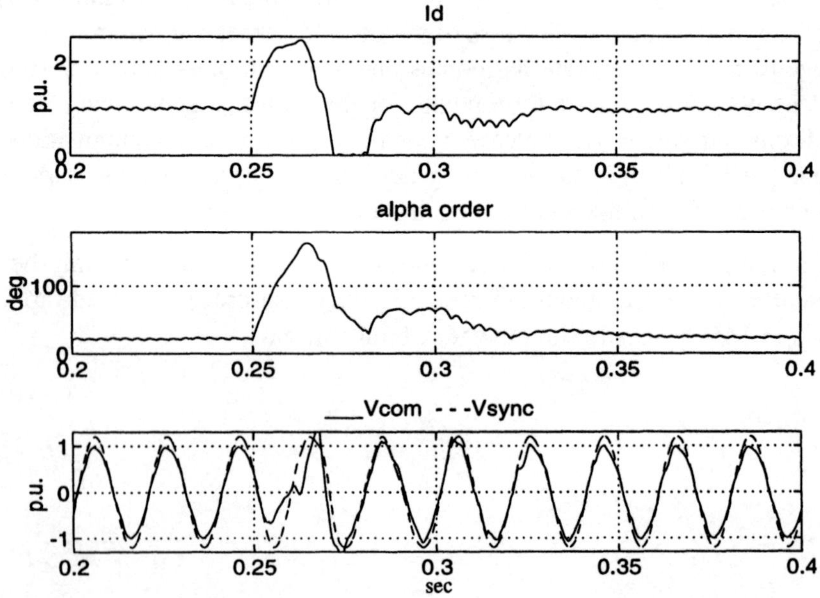

Figure 3-16b: DC line fault with DQO GFU

3.6 CONCLUSIONS

This chapter provided information and performance results of two typical GFUs used in practical HVDC systems. A performance comparison between the two GFUs showed that they are equally capable of operation with a weak ac system having high levels of pollution and harmonic distortion in the commutation voltage. On the other hand, due to the absence of a low-pass filter in its control loop, the DQO GFU is relatively easier to design and is somewhat faster than the conventional type.

In any application of a static converter in an utility environment, a means to synchronize the control system with the ac system plays a very important role. This design information presented here will be useful to utility planners for system studies.

3.7 ACKNOWLEDGEMENT

The contributions of Mr. V. Khatri are gratefully acknowledged.

3.8 REFERENCES

[1]. J. D. Ainsworth, "The Phase-Locked oscillator: A new control system for controlled static converters", IEEE Trans. Power App. System, Vol. PAS-87, pp 859-865, 1968.

[2]. J. D. Ainsworth, "Harmonic instability between controlled static converter and ac networks", Paper 5321, Proc. IEEE May, 1967.

[3]. A. Ekstorm and G. Liss, "A refined HVDC control system", IEEE Trans. on PAS-89, No. 5/6, May/June 1970, pp 723-732.

[4]. E. Rumpf and S. Ranade, "Comparison of suitable control schemes for HVDC stations connected to weak ac systems, Part I: new control systems", IEEE Trans. Power App. System, Vol. PAS-91, pp. 555-564, 1972.

[5]. D. H. Wolaver, "Phase locked Loop Circuit Design" Prentice Hall, Inc. 1991.

[6]. K. Bayer, H. Waldman and M. Weibelzahl, "Field Oriented Closed Loop Control of an Synchronous Machine with the New Transvektor Control System", Siemens Review XXXIX (1972), pp 220-223

[7]. M. Szechtman et al, "First benchmark model for HVDC control studies", Electra, April 1991, No.135, pp 55-73.

[8]. J. Jacques, F. P. Dawson and R. Bonert, "A tracking bandpass filter for the

rejection of power system disturbances", 1993 EPE'93 5th European Conf. on Power Electronics and Applications, IEE Conf. Publication No. 377, Vol. 4, 13-16 Sept. 1993, Brighton, UK. pp 47-52.

[9]. V. K. Sood, V. Khatri and H. Jin, "Performance assessment using EMTP of two GFUs for HVDC converters operating with weak ac systems", International Conference on Power System Transients, Sept. 3-7 1995, Technical University of Lisbon, Portugal. pp 517-522.

Chapter 4

HVDC Controls

The historical background to the developments that took place in the evolution of HVDC controllers will be presented in this chapter. The basis and formulations of modern controllers will be discussed. Because these controllers have an important bearing upon the interconnected AC system, a section will be devoted to inter-actions between the controllers and the system with examples drawn from practical systems.

4.1 HISTORICAL BACKGROUND

To control the firing angle of a converter, it is necessary to synchronize the firing pulses emanating from the trigger unit to the ac line commutation voltage which has a frequency of (60 or) 50 Hz in steady state. In the early 1950s, when the first HVDC converter installations were implemented with mercury arc valves, the relative size of the terminals was small compared to the MVA capacity of the ac systems coupled to these converters. This essentially meant that the grid firing system [1], which was synchronized directly to the sinusoidal ac system waveform, could generate the firing pulses in a relatively stable manner. However, since the three-phase sinusoidal ac waveforms of the ac systems were used as the synchronizing elements, the firing pulses were individually generated for each of the valves of the converter.

In the early 1960s, problems of firing pulse synchronization [2,3,4] were observed due to the distortion of the ac waveform caused by harmonic instability. It was noted that the commutation voltage was neither constant in frequency nor amplitude during a perturbed state. However, it is only the frequency that is of primary concern for the synchronization of firing pulses. For strong ac systems, the system frequency is relatively distortion free to be acceptable for most converter type applications. But, as converter connections to weak ac systems became required more often than not, it became necessary to devise a scheme for synchronization purposes which would be **decoupled** from the commutation voltage for durations when there were perturbations occurring on the ac system.

The most obvious method was to utilize an **independent oscillator** at (50 or) 60 Hz which could be **synchronously locked** to the ac commutation voltage. This oscillator would then provide the (phasor) reference relationship to the trigger unit during the perturbation periods, and would use the steady state periods for locking in step with the system frequency. The advantage of this independent oscillator was to provide an ideal (immunized and clean) sinusoid for synchronizing and timing purposes. Due to its timing stability, it offered the possibility of equi-distant firing pulses [5,7,13] which eliminated the generation of non-characteristic harmonics during steady-state operation. This was a prevalent and undesirable feature during the use of the earlier Individual Phase Control (IPC) system where the firing pulses were directly coupled to the commutation voltage, V_{com}.

There were two possibilities for this independent oscillator:

- Use of a fixed frequency oscillator (also called the Pulse Phase Control Oscillator (PPCO)) operating at a fixed frequency of 60 Hz. However, since the system frequency actually drifts between say 55-65 Hz due to the generators used to produce electricity, it was necessary to employ a control loop to track the drifting firing angle. One manufacturer employed a Current Controlled Oscillator (CCO) for this technique.
- Use of a variable frequency oscillator (also called the Pulse Frequency Control oscillator) with a locking range of between say 55-65 Hz and the centre frequency of 60 Hz. This oscillator would then employ a control loop of some sort for tracking the drifting system frequency; the control loop would have its own gain and time constants for steady state accuracy and dynamic performance requirements [19]. Two manufacturers employed a Voltage Controlled Oscillator (VCO) for this approach.

During the mid 1960s, industry, therefore, switched to this type of synchronization unit based on an independent, frequency controlled oscillator controlled by either a voltage or a current source. Both versions relied on an independently controlled oscillator whose frequency was decoupled from the frequency variations of the ac system feeding the converter. This meant that the converter firing pulses could now be truly equi-distant in the steady state.

With time, however, the variant with PPC [8] has become virtually obsolete, and only the PFC [5,7] variant is presently being used by industry.

The equi-distant pulse firing control systems used in modern HVDC control systems are identical to those developed in the mid-1960s [5,7]; although improvements have occurred in their hardware implementation since then, such as the use of micro-processor based equipment etc., their fundamental philosophy has not changed much.

4.2 FUNCTIONS OF HVDC CONTROLS

In a typical two-terminal dc link connecting two ac systems (Figure 4-1), the primary functions of the dc controls are to:

- Control power flow between the terminals,
- Protect the equipment against the current/voltage stresses caused by faults, and
- Stabilize the attached ac systems against any operational mode of the dc link.

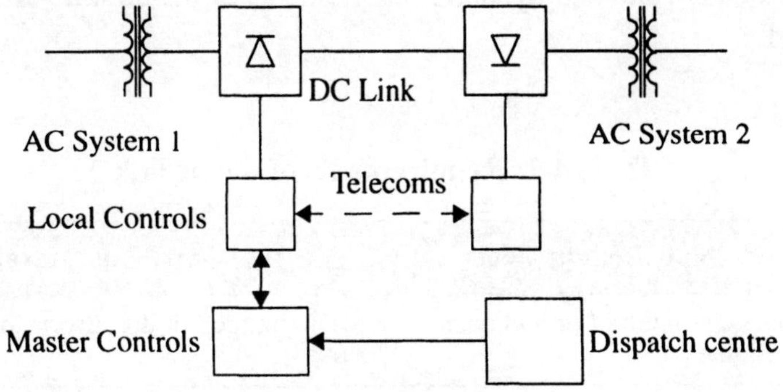

Figure 4-1: Typical HVDC system linking two ac systems

The two dc terminals each have their own local controllers. A centralized dispatch centre will communicate a power order to one of the terminals which will act as a Master Controller and has the responsibility to coordinate the control functions of the dc link. Besides the primary functions, it is desirable that the dc controls have the following features:

- **Limit the maximum dc current.**

  Due to a limited thermal inertia of the thyristor valves to sustain overcurrents, the maximum dc current is usually limited to less than 1.2 pu for a limited period of time.

- **Maintain a maximum dc voltage for transmission.**

  This reduces the transmission losses, and permits optimization of the valve rating and insulation.

- **Minimize reactive power consumption.**

  This implies that the converters must operate at a low firing angle. A typical converter will consume reactive power between 50-60% of its MW rating. This amount of reactive power supply can cost about 15% of the station cost, and consume about 10% of the power loss.

- **Other features.**

  Such as the control of frequency in an isolated ac system or enhance power system stability.

In addition to the above desired features, the dc controls will have to cope with the steady-state and dynamic requirements of the dc link, as shown in Table 4-1.

Table 4-1: Requirements of the dc link

Steady State Requirements	Dynamic Requirements
Limit the generation of non-characteristic harmonics	Step changes in dc current or power flow
Maintain the accuracy of the controlled variable, i.e. dc current and/or constant extinction angle	Start-up and fault induced transients
Cope with the normal variations in the ac system impedances due to topology changes	Reversal of power flow
	Variation in frequency of attached ac system

4.3 CONTROL BASICS FOR A TWO-TERMINAL DC LINK

A two terminal dc link is shown in Figure 4-2 with a rectifier and an inverter. The dc system is represented by an inductance L and a line resistance R; the value of the inductance L comprises the smoothing reactor(s), dc line inductance whereas the value of R includes the resistances of the smoothing reactor(s) and the resistance of the dc line etc.

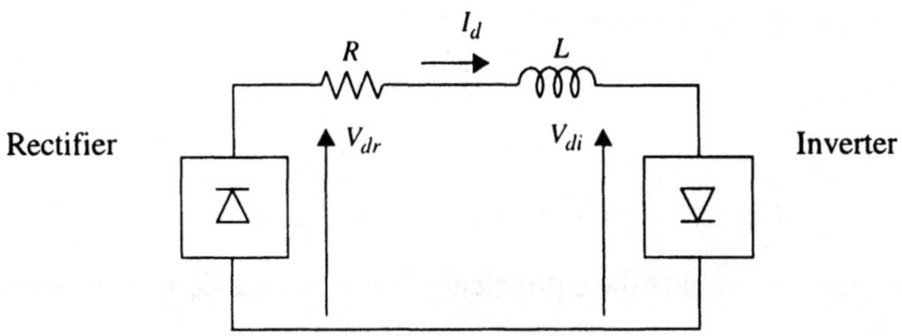

Figure 4-2: A two terminal dc link

Using Ohm's law, the dc current I_d in the dc link depicted in the figure is given as

$$I_d = (V_{dr} - V_{di})/R \tag{4-1}$$

where V_{dr} - dc voltage of the rectifier,

V_{di} - dc voltage of the inverter, and

R - is the dc line resistance.

The power flow transmission of the dc link is therefore given by

$$P_d = V_d * I_d \tag{4-2}$$

From converter theory, in the case of a CSC the V_d-I_d relationship for a rectifier is given by:

$$V_{dr} = V_{dor} Cos\alpha - R_{cr} I_d \quad (4\text{-}3)$$

From converter theory, in the case of a CSC the V_d-I_d relationship for an inverter is given by either

$$V_{di} = V_{dir} Cos\beta - R_{ir} I_d \quad (4\text{-}4)$$

or, depending on choice of control variable

$$V_{di} = V_{dir} Cos\gamma - R_{ir} I_d \quad (4\text{-}5)$$

Using equations describing V_{dr} and V_{di} for the case of a CSC, the dc line current is given by either one of two options depending upon the choice of the control mode at the inverter:

$$I_d = (V_{dor} Cos\ \alpha \ - \ V_{doi} Cos\ \beta) / (R + R_{cr} + R_{ci}) \quad (4\text{-}6)$$

or

$$I_d = (V_{dor} Cos\ \alpha \ - \ V_{doi} Cos\ \gamma) / (R + R_{cr} - R_{ci}) \quad (4\text{-}7)$$

These equations provide the equivalent circuits for the dc link, as shown in Figure 4-3.

Changes in I_d can therefore occur by:

1. Varying α at the rectifier. Due to electronic control, this is quite fast and will occur within one-half cycle (or about 8-10 milli-seconds).
2. Varying β or γ at the inverter. This is quite fast and will occur within milli-seconds.
3. Varying AC voltage at rectifier by means of the converter transformer tap changer. This is a slow process and usually takes the order of several hundreds of seconds.
4. Varying AC voltage at inverter by means of the transformer tap changer. This is a slow process and usually takes the order of several hundreds of seconds.

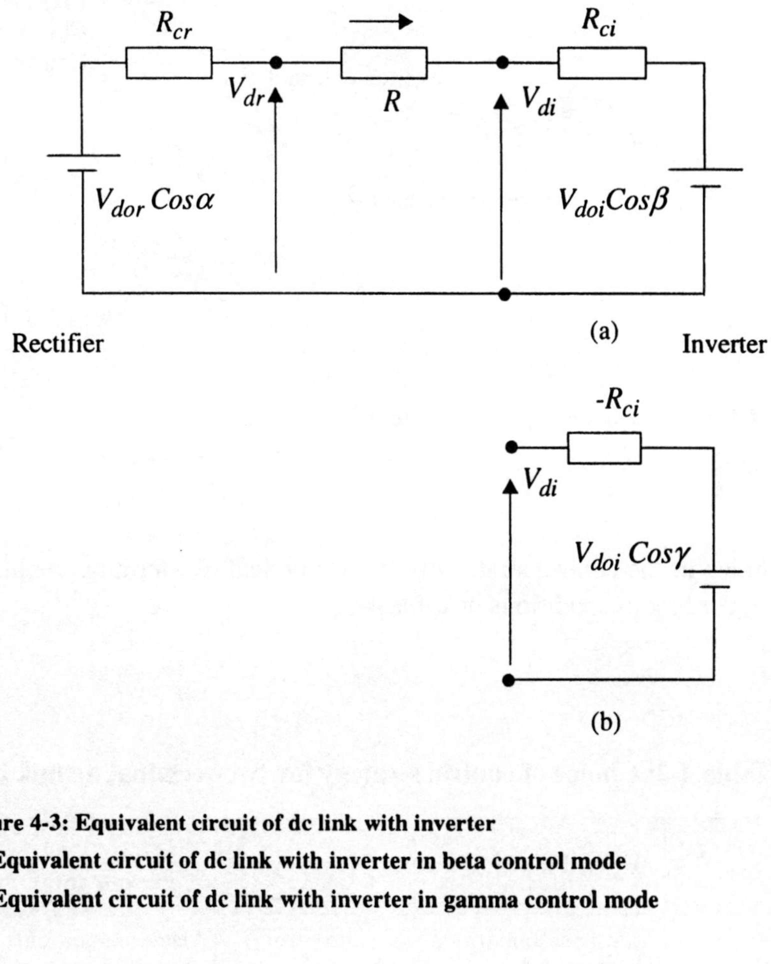

Figure 4-3: Equivalent circuit of dc link with inverter
(a) Equivalent circuit of dc link with inverter in beta control mode
(b) Equivalent circuit of dc link with inverter in gamma control mode

The choice of control strategy is selected to enable a fast and stable operation of the dc link whilst minimizing the generation of harmonics, reactive power consumption and power transmission losses.

The three characteristics represent straight lines on the V_d-I_d plane, as shown in Figure 4-4. Notice that eq. (4-4), i.e. the beta characteristic, has a positive slope while the eq. (4-5), i.e. the gamma characteristic, has a negative slope.

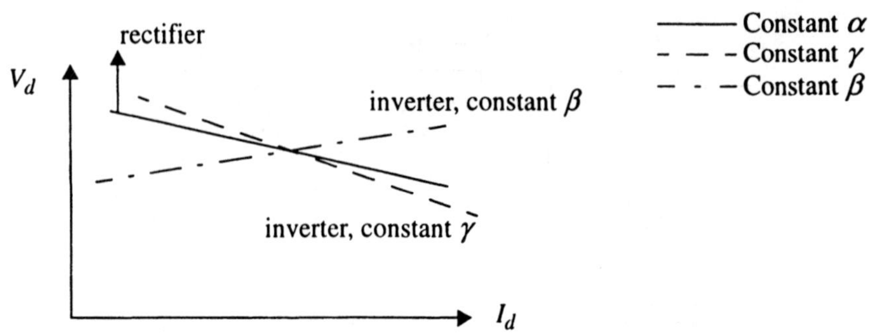

Figure 4-4: Representation of control strategy

The choice of the control strategy [12] for a typical two-terminal dc link is made according to conditions in Table 4-2.

Table 4-2: Choice of control strategy for two-terminal dc link

Condition #	Desirable features	Reason	Control implementation
1	Limit the maximum dc current, I_d	For the protection of valves	Use constant current control at the rectifier
2	Employ the maximum dc voltage, V_d	For reducing power transmission losses	Use constant voltage control at the inverter
3	Reduce the incidence of commutation failures	For stability purposes	Use minimum extinction angle control at inverter
4	Reduce reactive power consumption at the converters	For voltage regulation and economic reasons	Use minimum firing angles

Condition 1 implies the use of the rectifier in constant current control mode and Condition 3 implies the use of the inverter in constant extinction angle (CEA) control mode. Other control modes may be used to enhance the power transmission during contingency conditions depending upon applications.

4.4 CURRENT MARGIN CONTROL METHOD

The so called "Current Margin Method" of control for two terminal HVDC system is the most widely accepted method in use at present. The method relies on a defined zone of operation of the dc system, with clear functions for both terminals. It also incorporates protection features to protect the dc link [12].

4.4.1 Rectifier mode of operation

The rectifier mode of operation is defined by a number of characteristics as shown in the Figure 4-5.

1. Alpha-min characteristic at rectifier

From converter theory, it can be shown that

$$V_d = V_{dor} \cdot \cos\alpha - R_{cr} \cdot I_d \tag{4-8}$$

where $R_{cr} = \frac{3}{\pi} \cdot w \cdot L_{cr}$

Equation (4-8) describes a straight line AB when plotted as a V_d-I_d characteristic in steady state, as shown in Figure 4-5. The slope of this characteristic is the value $-R_{cr}$ which is defined as the equivalent commutation resistance; a low value of R_{cr} would imply a strong ac system, and the characteristic would be almost horizontal. The intercept of this characteristic on the V_d axis is equal to the value $V_{dor} Cos\alpha$ when the value of $I_d = 0$. The maximum limit of the voltage V_d will be defined by the value of $\alpha = 0$ degs., i.e. when the rectifier is a theoretical diode converter with firing angle equal to zero. In reality, a minimum value of about α = 2-5 degs. is normally required to ensure that the converter valves have a minimum positive voltage for turning on. This zone is bounded by the hatched area in Figure 4-5.

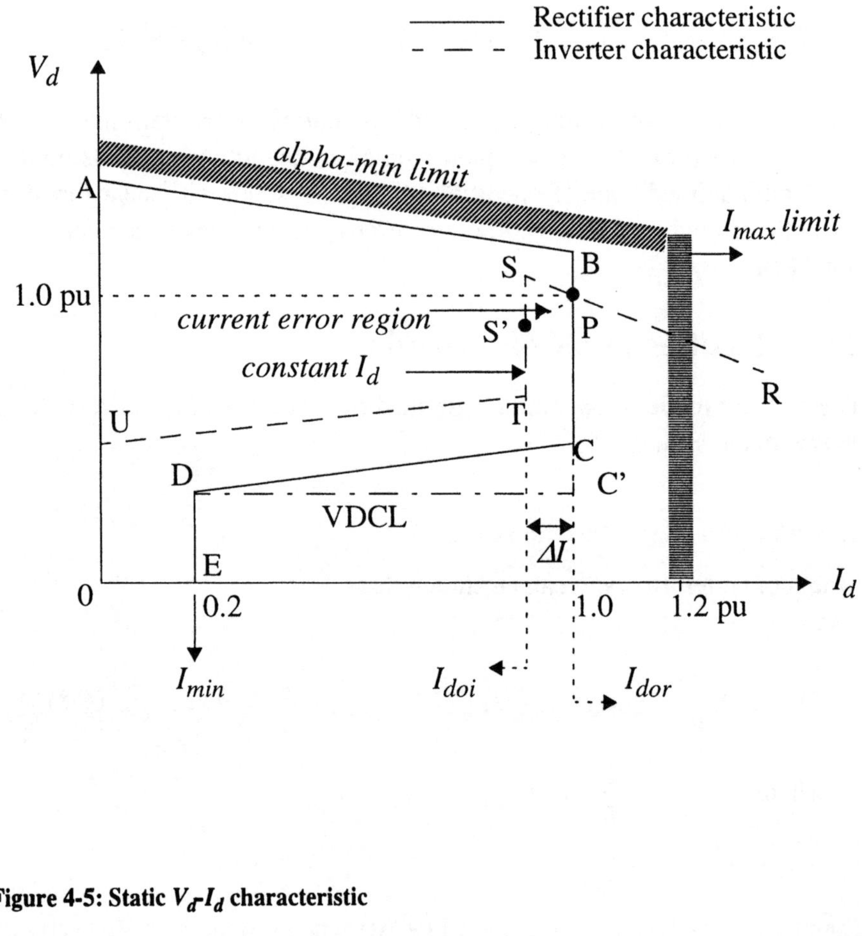

Figure 4-5: Static V_d-I_d characteristic

2. Constant I_d characteristic:

The converter valves have limited thermal inertia, and therefore cannot carry a large current over their rated value for any extended period of time. Typically, a maximum limit of I_{max} =1.2 pu is specified as the upper limit for the current carrying capacity of the valves. The I_d = constant current characteristic is a straight line BC, as shown in the figure. The zone of operation is also indicated.

3. VDCL characteristic:

The Voltage Dependent Current Limit (VDCL) is a limitation imposed by the ability of the ac system to sustain the dc power flow when the ac voltage at the rectifier bus is reduced due to some perturbation. Some variants of this characteristic utilize a horizontal portion, as defined by C'D, instead of a sloped portion (as defined by CD).

4. I_{min} characteristic:

This limit is usually imposed to maintain enough dc current in the valves to avoid reaching discontinuous current operation with its consequential current chopping phenomena which could lead to dangerous transient dc voltages. Typical values of I_{min} are between 0.2-0.3 pu.

4.4.2 Inverter Mode of operation

1. Gamma-min characteristic:

Equation (4-9) defines the V_d-I_d characteristic at the inverter; although there are two possibilities, the minimum extinction angle (gamma) option is utilized generally. The line SR (Figure 4-5) defines this mode of operation and is referred to as the Constant Extinction Angle (CEA) control mode. The slope of this line is usually more pronounced than the corresponding one for the rectifier due to the relative strength of the inverter-end ac system.

$$V_d = V_{doi} \cdot \cos\gamma - R_{ci} \cdot I_d \tag{4-9}$$

where $R_{ci} = \frac{3}{\pi} \cdot w \cdot L_{ci}$

2. Constant current characteristic:

The line ST defines the Constant Current characteristic of operation at the inverter. In order to maintain a unique operating point of the dc link, defined by the cross-over point P of the characteristics of the rectifier and inverter, a current margin of $\Delta I_d = 0.1$ pu is normal for the current orders given to the rectifier (I_{dor}) and inverter (I_{doi}) i.e. I_{dor} - I_{doi} = ΔI_d.

However, the current demanded by the inverter I_{di} is usually less than the current demanded by the rectifier I_{dr} by the current margin ΔI which is typically about 0.1 pu; its magnitude is selected to be large enough so that the rectifier and inverter constant current modes do not interact due to any current harmonics which may be superimposed on the dc current. This control strategy is termed the current margin method.

3. Alpha-min in inverter mode:

The line TU defines the alpha-min-in-inverter mode characteristic. This value is typically about 100-110 degrees, and is required to limit any excursions (even transiently) of the inverter into the rectifier mode of operation. Furthermore, the value of 100-110 degrees ensures a minimum dc voltage at the inverter during a fast start-up of the dc link with $I_d = 0$.

4. Current error region:

A modification to the inverter characteristic (line PS' in Figure 4-5) is often made during the current margin period to avoid any instability due to multiple operating points occurring with a weak ac system at the inverter.

This modification is illustrated in Figures 4-6a and b. Note that the VDCL characteristic has been eliminated in these figures for reasons of simplicity.

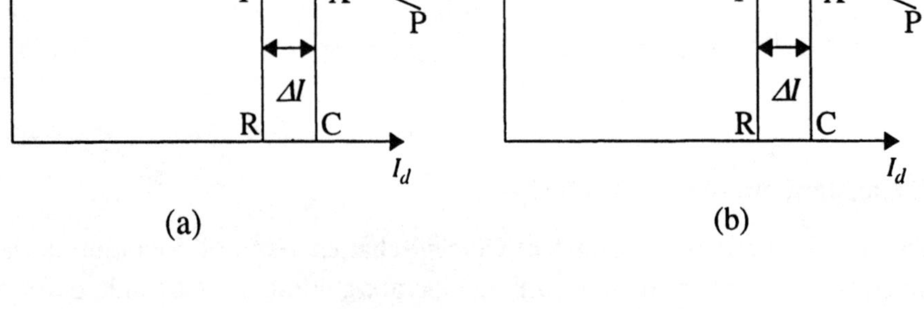

Figure 4-6: Static V_d-I_d characteristic for a two-terminal link (a) unmodified and (b) modified

The rectifier characteristic is composed of two control modes: alpha-min (line AB) and constant-current (line BC). The alpha-min mode of control at the rectifier is imposed by the natural characteristics of the rectifier ac system and the ability of the valves to operate at an alpha angle equal to zero i.e. in the limit the rectifier acts a diode rectifier. However, since a minimum positive voltage is desired before firing of the valves to ensure conduction, an alpha-min limit of about 2-5 degrees is normally imposed.

The inverter characteristic is composed of two modes: gamma-min (line PQ) and constant-current (line QR). The operating point for the dc link is defined by the cross-over point X of the rectifier and inverter characteristics. In addition, a constant current characteristic is also used at the inverter. However, the current demanded by the inverter I_{di} is usually less than the current demanded by the rectifier I_{dr} by the current margin ΔI which is typically about 0.1 pu; its magnitude is selected to be large enough so that the rectifier and inverter constant current modes do not interact due to any current harmonics which may be superimposed on the dc current. This control strategy is termed the current margin method.

The advantage of this control strategy becomes evident if there is a voltage decrease at the rectifier ac bus forcing the line AB to move downwards. The operating point then moves to point Y and the inverter takes over current control. This way the current transmitted will be reduced to 0.9 pu of its previous value and voltage control will shift to the rectifier. However, the power transmission will be largely maintained near to 0.9 pu of its original value.

The control strategy usually employs the following other modifications to improve the behavior during system disturbances (Figure 4-7):

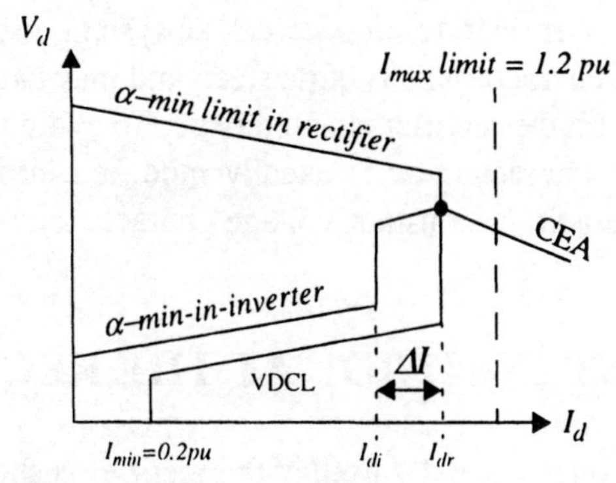

Figure 4-7: Complete static V_d-I_d characteristic for a two terminal HVDC system

At the rectifier:

1. Voltage Dependent Current Limit, VDCL

This modification is made to limit the dc current as a function of either the dc voltage or, in some cases, the ac voltage. This modification assists the dc link to recover from faults. Variants of this type of VDCL do exist. In one variant, the modification is a simple fixed value instead of a sloped line.

2. I_d-min limit

This limitation (typically 0.2-0.3 pu) is to ensure a minimum dc current to avoid the possibility of dc current extinction caused by the valve current dropping below the hold-on current of the thyristors; an eventuality that could arise transiently due to harmonics superimposed on the low value of dc current. The resultant current chopping would cause high overvoltages to appear on the valves. The magnitude of I_d-min is affected by the size of the smoothing reactor employed.

At the inverter:

1. Alpha-min limit at inverter

The inverter is usually not permitted to operate inadvertently in the rectifier region, i.e. a power reversal occurring due to, say, an inadvertent current margin sign change. To ensure this, an alpha-minimum-limit in inverter mode of about 100-110 degrees is imposed.

2. Current error region

When the inverter operates into a weak ac system, the slope of the CEA control mode characteristic is quite steep and may cause multiple crossover points with the rectifier characteristic. To avoid this possibility, the inverter CEA characteristic is usually modified into either a constant Beta characteristic or constant voltage characteristic within the current error region.

4.5 CURRENT CONTROL AT THE RECTIFIER

The current controller normally used at the rectifier is shown in Figure 4-8. A measurement of the dc current of the dc system is obtained and compared to a reference value I_o. The resultant current error I_e is then fed to the PI regulator with proportional and integral gains K_p and K_i respectively. The output of this regulator is a voltage signal known as the alpha order α_o which controls the frequency output of the VCO. Under conditions of α_o= 0, the output frequency of the VCO is a constant at 360 Hz and is set by the input U_{ref} which could be a voltage proportional to a measurement of the ac system frequency; this way any slow modulations of the system frequency could also be compensated for, if required. The sum of α_o and U_{ref} is fed to the VCO which comprises of a resettable integrator and a comparator. The gain of the integrator is selected to be 720*Pi. The output of the comparator is a pulse train at 360 Hz, i.e. 6 times the fundamental frequency. The Ring Counter then derives individual firing pulses for the 6-pulse converter from the output pulse train of the comparator. It is noteworthy that the VCO presents an integral transfer function within the control loop.

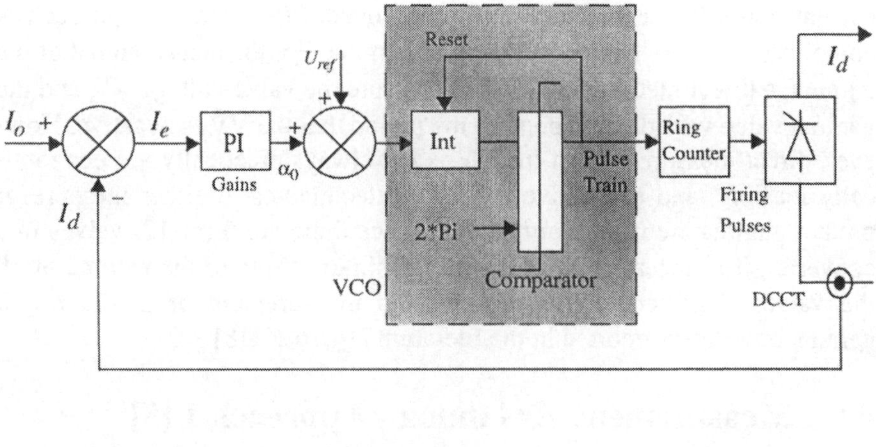

Figure 4-8: Current control at the rectifier end

The frequency of the VCO is at 6 times the fundamental frequency of the ac system. Yet the VCO is totally independent of the frequency of the ac system, other than having the locking or centre-frequency set at 6 times the fundamental frequency. The VCO frequency is therefore free to drift within a frequency range of f_o +/- Δf, where Δf is equal to $(f_{max} - f_{min})$. Typically

$f_{max} = 1.2*f_o$ and $f_{min} = 0.8*f_o$. The frequencies f_{max} and f_{min} represent the upper and lower frequency limits respectively of operation for the VCO.

The mechanism for controlling the frequency of the output of the VCO will now be explained. The negative feedback, current-error loop provides a method to synchronize the independent VCO output frequency with the ac system frequency. Any deviation of the current-error signal from zero will either speed up or slow down the VCO to maintain synchronism with the ac system frequency. For this reason, the term "phase locked oscillator" was coined by its inventor [5]. (In the case of an inverter, an alternative negative feedback loop utilizing constant extinction angle would provide a similar means to synchronize the VCO frequency with the ac system frequency).

4.6 INVERTER EXTINCTION ANGLE CONTROL

For the extinction angle control for the inverter, a technique similar to the current controller at the rectifier is employed. However, the approach is complicated due to the measurement of gamma. For the measurement of the gamma, a direct method would be to measure the valve voltage VV, and the gamma value would correspond to the period that the VV is negative. However, direct measurement of the VV is not always practically nor economically feasible, and alternative or indirect techniques to either measure or predict gamma are used. Furthermore, since there are 6 (or 12) valves in a converter, it is necessary to obtain the minimum value of the gamma of all the valves. Different approaches for the measurement or prediction of gamma have been reported in the literature [14,16,17,18].

4.6.1 Measurement of Gamma - Approach 1 [5]

One method uses the moment of the firing of the out-going valve and the detection of current zero in that valve to determine the value of the overlap angle μ (Figures 4-9 and 4-10). The ac commutation voltage zero cross-over point, with the voltage going positive, then provides the end of the gamma angle γ. Hence, the ignition angle β can be calculated from a knowledge of the period from the moment of firing of the out-going valve to the moment of the commutation voltage reversal, going positive i.e. $\beta = \mu + \gamma$.

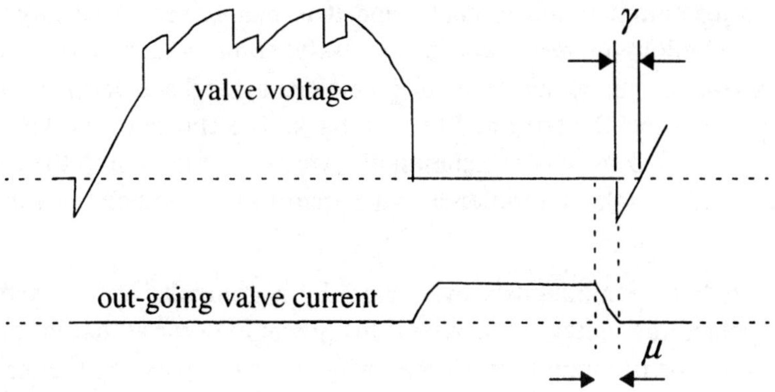

Figure 4-9: Measurement of gamma - approach 1

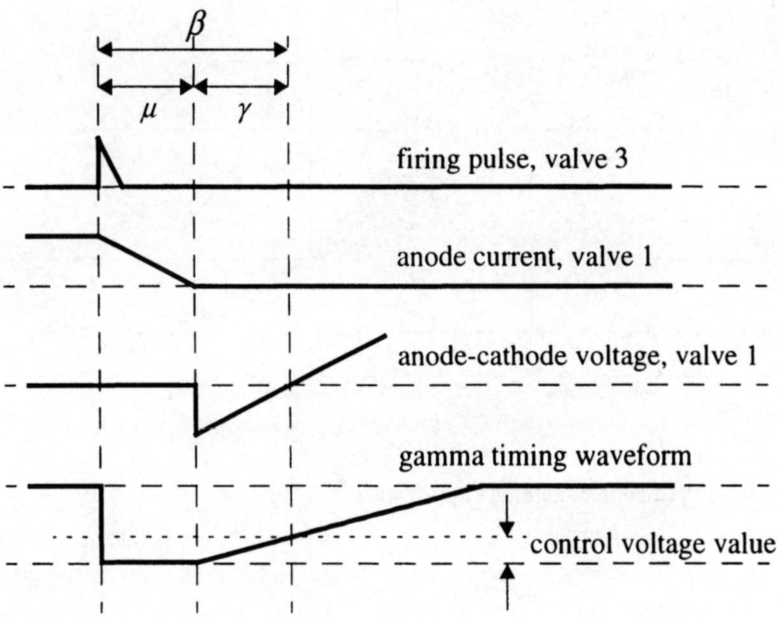

Figure 4-10: Measurement of gamma - approach 1

4.6.2 Prediction of Gamma - Approach 2 [7]

In this method, a prediction of the remaining commutation voltage-time area after commutation is made, and it is maintained to be larger than a specified minimum necessary for successful commutation. The prediction is approximate, but to increase its precision, a feedback loop is employed which measures the error and feeds it back. The choice of the voltage-time area is justified since commutation of a valve is a function of the remaining commutation voltage-time area rather than just the remaining time period alone.

The predictor continuously calculates (by a triangular approximation) the total remaining voltage-time area if firing would occur at that instant. Since the predictor is common to all the valves in one 6-pulse converter, it operates for a period of 60 degrees per valve. Figure 4-11 shows this function for one commutation voltage U_{sr} for valve 3 of the converter bridge.

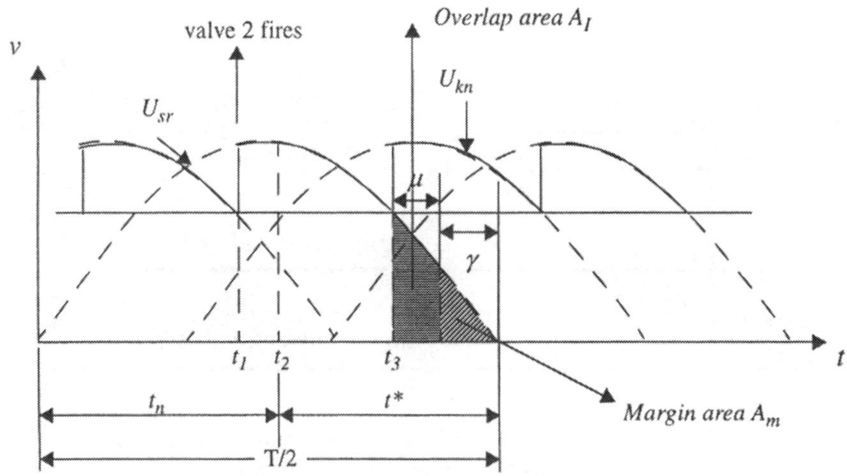

Figure 4-11: Prediction of gamma - approach 2

At instant t_1, valve 2 fires and a special selector circuit connects the voltage U_{sr} to the predictor circuit. If firing were to occur at instant t_2, the remaining voltage-time area after finished commutation will be:

$$A_m(t_2) = \int_{(t-t_2)}^{T/2} (U_{sr} \cdot \sin wt)dt - (k \cdot I_d) \tag{4-10}$$

The term kI_d in the above equation makes an allowance for the fact that the overlap commutation voltage-time area is directly proportional to the dc current I_d. The integral term in the equation can be approximated by a triangular area:

$$A_p(t_2) = \frac{U_{kn}(t) \bullet t^*(t_2)}{2} \tag{4-11}$$

where t^* is the predicted remaining time to the next zero crossing of the commutation voltage; it is calculated as the difference between the measured value of $T/2$ and the period t_n. i.e. $t^* = T/2 - t_n$. The period t_n started from the previous zero crossing of the voltage U_{sr} (as can be seen in Figure 4-11).

The voltage t_n is generated in a three phase device from which the actual phase is selected at each firing instant and connected to the predictor. This is illustrated in the Figure 4-12.

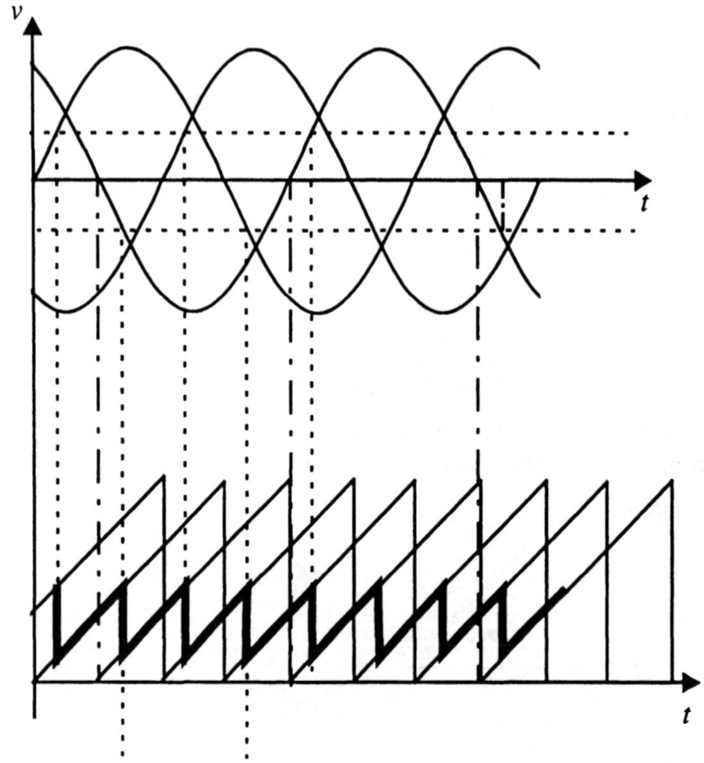

Figure 4-12: Prediction of gamma

The prediction process follows the relationship

$$A_{m} \cdot {}_{pred}(t) = U_{kn}(t) \cdot t^{*}(t) - kI_{d} \tag{4-12}$$

The approximate method of prediction gives reasonably good values at low values of dc currents with a sinusoidal commutation voltage. However, with large values of dc current and overlap angle, or with harmonic distortion of the commutation voltage, the error becomes significant enough to merit correction through a feedback loop. To achieve this the value of the predicted gamma is stored in a holding circuit. The actual gamma is measured (one cycle later) and subtracted from the previously held value of gamma. The prediction error is defined as:

$$\Delta A = A_{m\ pred} - A_{m\ \text{real}} \tag{4-13}$$

where $A_{m\ real}$ is the actual margin. After filtering, ΔA is used as a correction signal for the reference value one period (cycle) later. Thus the complete expression for the firing condition is defined as:

$$(A_{m\ pred})_k = A_{m\ ref} + (\Delta A)_{k-1} \tag{4-14}$$

where the index k indicates the instant k and the index $(k\text{-}1)$ the corresponding point one period later.

The prediction process has an inherent individual phase character. If no further action were taken, each valve would fire on the minimum area margin condition. To counteract this a special firing symmetrizer is used. When one valve has fired on the minimum margin condition, the following five valves are then fired equidistantly by means of the voltage controlled oscillator.

Once a measurement of the gamma angles from the six valves of the convertor are obtained, the minimum value is selected. This value is then compared to the desired value of gamma and an error signal is generated and fed to a PI regulator. This gamma error signal is used in a similar manner to the current controller at the rectifier to generate the firing pulses for the converter.

4.7 HIERARCHY OF CONTROLS

The terminal at one end of the DC transmission system is shown in the Figure 4-13. The terminal can be divided into sub-sections i.e. a Bipole which comprises of the positive and negative poles. Each pole can be further subdivided into the star valve group and the delta valve group depending on the transformer configuration used. Each valve group comprises a 6-pulse converter.

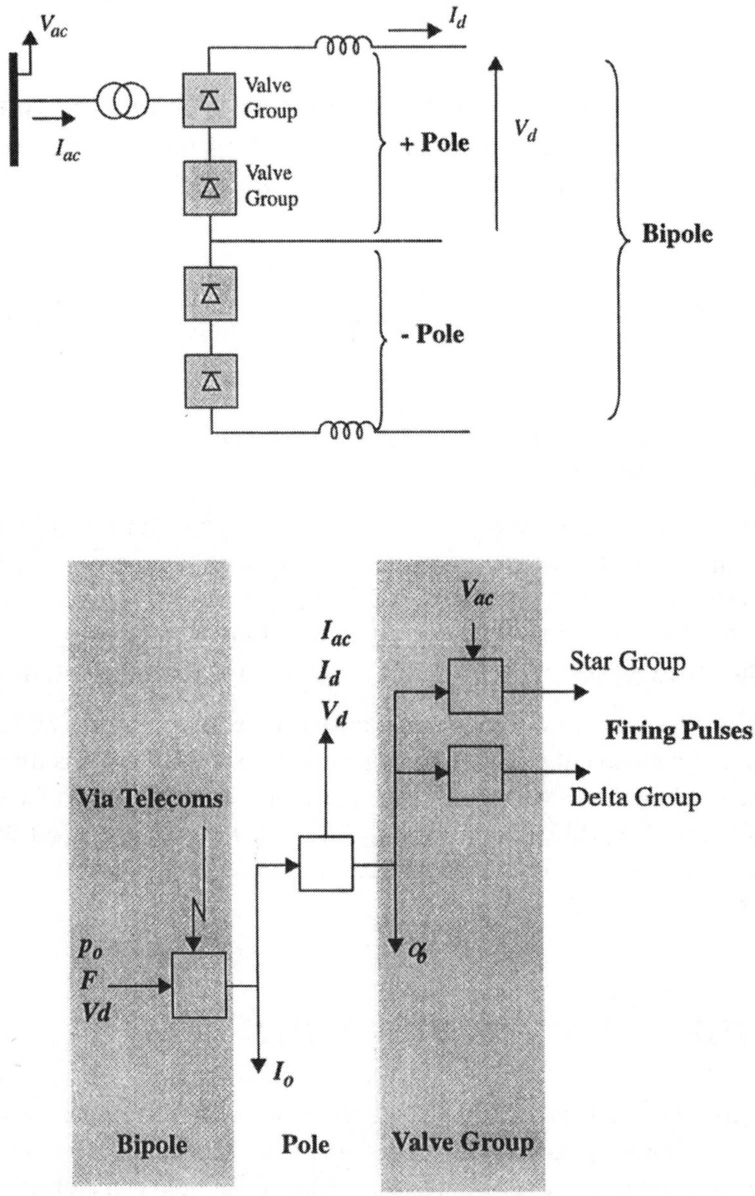

Figure 4-13: Hierarchy of controls

Details of the controllers at different hierarchical levels of the terminal are provided next.

4.7.1 Bipole Controller (Figure 4-14)

The bipolar controller usually receives a power order Po from the station operator. This is normally subjected to a controlled rate of increase/decrease in order to protect the system from sudden changes in desired power. A supplementary power modulation signal ΔP_o can also be inputted at this stage, if required. Maximum power P_{max} and minimum power P_{min} limits to the excursions of the power controller are imposed. Finally, the power order is divided by the DC voltage measured value to derive a current order I_o which is sent to the two Pole Controllers. For this circuit, in case of start up routine when the dc voltage may be zero or low value, a bias circuit is required to counteract any problem due to a divide-by-zero function. The output of this controller is the I_o limited value which is subjected to the protective Voltage Dependent Current Limit (VDCL).

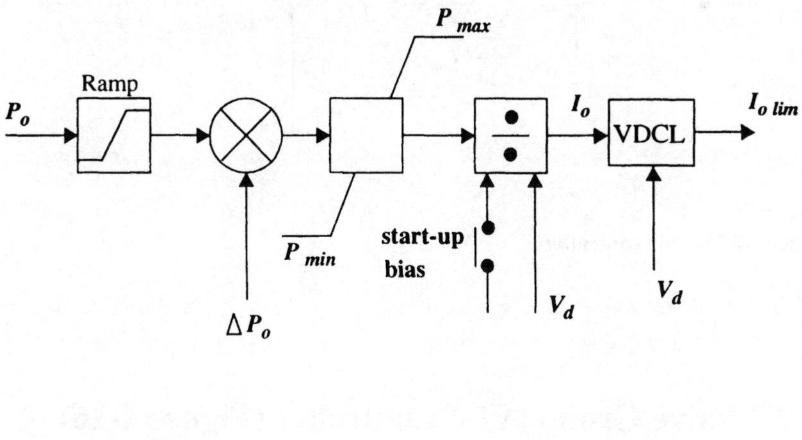

Figure 4-14: Bipole controller

4.7.2 Pole Controller (Figure 4-15)

The input to each of the Pole Controllers is the current order I_o from the Bipole Controller. The supplementary current input ΔI_o can be added to this I_o to achieve any modulation of the order if desired. The current input is subjected to upper I_{max} and lower I_{min} limits for protective purposes. After limitation, the current order is compared to the measured value of dc current I_d to generate an error signal I_e. Another signal which modifies the current order is the current margin ΔI which is required only at the inverter end to bias off the current controller so that the gamma controller can take over. The current controller uses the PI regulator to provide dynamic properties to the control loop, and provides the alpha order α_o at its output.

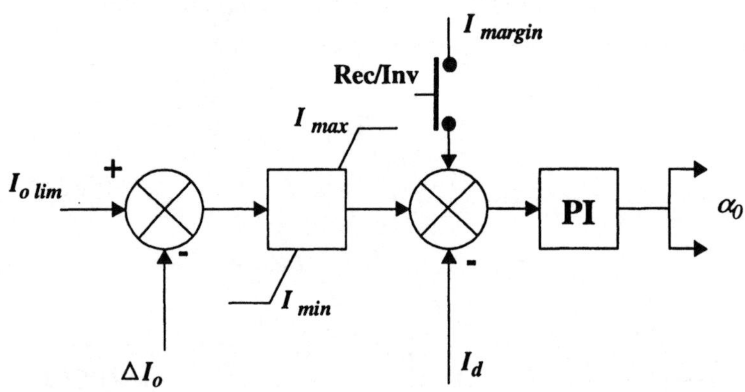

Figure 4-15: Pole controller

4.7.3 Valve Group (VG) Controller (Figure 4-16)

The alpha order signal from the pole controls is used to generate the firing pulses for the converter in the valve group controller. The VG controller has two separate secondary loops associated with it:

1. Tap Changer (TC) Controller

This is a relatively slow-acting loop (time constant of the order of several hundreds of milli-seconds) which maintains the tap position of the converter transformer. Its function is to maintain the firing angle alpha within a nominal range of about 15 degrees whenever it hits any limits by either raising or lowering the tap position. This will then minimize the reactive power consumption of the converter, and provide sufficient margin for dynamic operation of the converter.

2. The Commutation Failure (CF) Controller

This loop detects the possibility of a CF from measurements of the ac current, commutation voltage and the dc current. Rapid pre-programmed changes to the alpha order can be made as a function of the CF detector for assisting the recovery of the dc system from a CF.

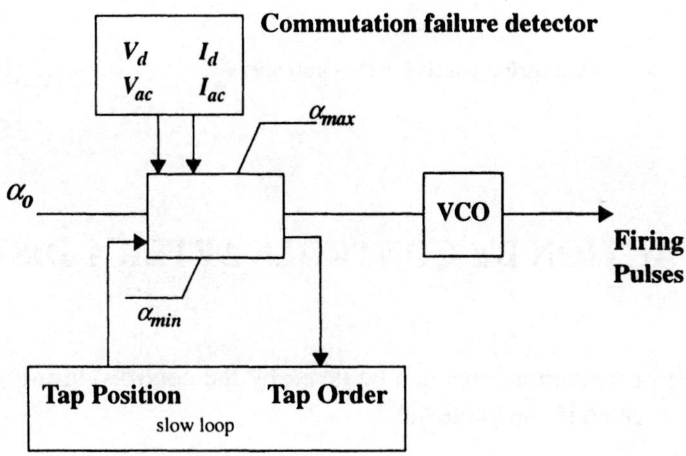

Figure 4-16: Valve group controller

The measurements available for control and protection purposes comprise the 3-phase commutation ac voltages V_{ac}, the ac currents I_{ac} on the primary and secondary sides of the converter transformer, and the dc current I_d and dc voltage V_d. The other input/output variables from the various controls are depicted in the Figure 4-17.

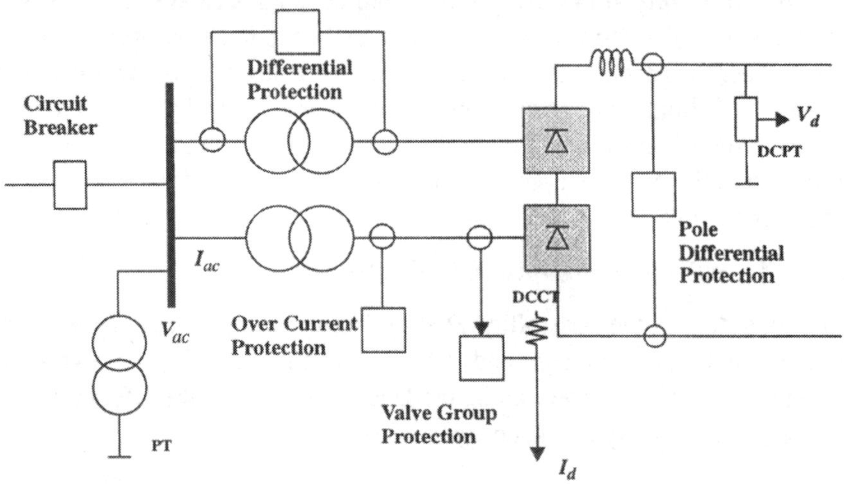

Figure 4-17: Monitoring points for the controllers

4.8 ACTION BY CONTROLS AFTER A DISTURBANCE

Some of the actions that can be taken by the controls during a disturbance are indicated in the Table 4.3.

Table 4-3: Control actions after disturbances

Disturbance	Action
Phase Distortion	$\beta_{minimum}$ limit increased temporarily from 150°-60° for 100 ms
Rectifier AC faults	$\alpha_{minimum}$ limit increased temporarily from 5° - 45°
Inverter AC faults - 1 Commutation Failure - n Commutation Failures	 γ - Increased transiently γ - Increased in stages
Block/Deblock or Restart	$\alpha_{minimum}$ limit increased to 60° and released slowly

4.9 REFERENCES

[1]. H.Forssel, "The Gotland d.c. link: The grid control and regulation equipment", Direct Current, 1955, 2, pp. 109-114.

[2]. P.G. Engstrom, "Operation and Control of HVDC Transmission," IEEE Trans. on Power Apparatus and Systems, Vol. 83, Jan. 1964, pp. 71-77.

[3]. T. Machida and Y. Yoshida, "A method to detect the deionization margin angle and to prevent the commutation failure of an inverter for dc transmission," IEEE Trans. on Power Apparatus and Systems, Vol. PAS-86, March 1967, pp. 259-262.

[4]. J.D. Ainsworth, "Harmonic instability between controlled static converters and a.c. networks", Proc. IEE, Vol. 114, pp 949-957, 1967.

[5]. J. Ainsworth, "The Phase-Locked Oscillator - A New Control System for controlled static converters", IEEE Trans. on Power Apparatus and Systems, Vol. PAS-87, No. 3, March 1968, pp.859-865.

[6]. N. Hingorani and P. Chadwick, "A new constant extinction angle control for ac/dc/ac static converters," IEEE Trans. on Power Apparatus and Systems, Vol. PAS-87, March 1968, pp. 866-872.

[7]. A. Ekstrom and G. Liss, "A Refined HVDC Control System", IEEE Trans. on Power Apparatus and Systems, Vol. PAS-89, No. 5/6, May/June 1970, pp. 723-732.

[8]. E. Rumpf and S. Ranade, "Comparison of Suitable Control Systems for HVDC Stations connected to weak ac system - Part I: New Control System," Part II: Operational Behaviour of the HVDC Transmission. IEEE Trans. on Power Apparatus and Systems, Vol. 91, Paper No. 71 TP 640 PWR, 1972, pp. 549-564.

[9]. A. Ekstrom, P. Jackson and G. Liss, "Firing Control if thyristor Valves in HVDC Transmissions", IFAC Symposium, Dusseldorf, Oct. 1974, pp. 509-525.

[10]. F. Nishimura, A.Watanabe, N.Fujii and F.Ogata, "Constant Power Factor Control System for HVDC Transmission", IEEE Trans. on Power Apparatus and Systems, Vol. PAS-95, No.6, Nov/Dec 1976.

[11]. F. Kay, "Investigation of Control problems during HVDC inverter operation," Proceedings of the 2nd IFAC Symposium on Control in Power Electronics and Drives, Dusseldorf, 1977.

[12]. R. Jotten et el, "Control of HVDC Systems - The State of the art. Part 1: Two Terminal Systems," CIGRE SC 14-10, Aug. 30 - Sept. 7, 1978.

[13]. J.D. Ainsworth, "Developments in the phase locked oscillator control system for HVDC and other large converters", IEE Conf. Publication 255 on AC and DC Power Transmission, Sept 1985, pp 98-103.

[14]. S.Bhattacharya and H.Dommel, "A new commutation margin control representation for digital simulation of HVDC system transients", IEEE Trans. on Power Systems, Vol.3, No.3, 1988, pp 1127-

[15]. Commutation Failures in HVDC Transmission Systems due to AC System faults, CIGRE Working Group 14.02, Electra No. 169, December 1996, pp 59-85.

[16]. M.Sato, K.Yamaji, M.Sekita et al, "Development of a Hybrid Margin Angle Controller for HVDC Continuous Operation", Paper 96 - WM 199-0 PWRS.

[17]. T.Funaki, and K.Matsuura, "Predictive Firing angle calculation for constant effective margin angle control of CCC-HVdc", Paper PE-320-PRD (03-2000).

[18]. F.Karlecik-Maier, "A new closed loop control method for HVDC transmission", IEEE Trans. on Power Delivery, Vol. 11, No,4, October 1996 pp. 1955-1960.

[19]. D.Jovcic, "Phase Locked Loop System for FACTS", IEEE Trans. on Power Systems, Vol. 18, No.3, Aug. 2003, pp1116 - 1124.

Chapter 5

Forced Commutated HVDC Converters

5.1 INTRODUCTION

Since the implementation of the first HVDC thyristor valve at Eel River (Canada) in the early 1970's, HVDC transmission has experienced tremendous growth. As power systems grow and become more integrated, interconnections to neighboring ac systems are becoming increasingly necessary to enhance stability, security of supply, flexibility and for other economic benefits. Primarily for stability reasons, the trend is for such interconnections to be asynchronous HVDC ties. And usually these interconnections feed into locations where the ac power systems are weak. Utility system planners realized that the critical element in the HVDC inter-tie was the thyristor converter, which has a fundamental limitation that it requires a reliable and adequately stiff voltage source for valve commutation purposes.

The traditional yardstick for assessment of the quality of this commutation voltage has been the Effective Short Circuit Ratio (ESCR) at the converter ac bus. For instance, a typical ac system with an approximate value of $ESCR \geq 2$ is considered adequate for a Line Commutated Converter (LCC) with some enhanced control techniques; systems with values of $ESCR < 2$ are considered as weak, and may suffer serious disruption to power transmission following a system disturbance, resulting in stability problems.

To overcome the requirement of adequate voltage source, forced commutation can be utilized to switch off the valves. The advantages/disadvantages of employing forced commutation are presented in Table 5-1:

Table 5-1: Advantages/disadvantages of forced commutation

Advantages	Disadvantages
• Ability to feed into weak or even dead ac systems, • Improved voltage regulation by control of active/reactive power flow, • Ability to recover rapidly (within 1 cycle) from system disturbances.	• Requirement for a commutation circuit [6-10], • Increased cost/kW for the converter, • Increased stresses on valves.

Although these advantages have been known for a long time, the disadvantages consistently blocked any serious applications of the technique until the early 1990s.

5.2 COMMUTATION TECHNIQUES FOR HVDC CONVERTERS

Before describing commutation techniques, some terminology is presented to avoid any confusion.

5.2.1 Definition Of Commutation

Within the context of HVDC converters, the definition of commutation is the transfer of dc current (I_d) from one valve to another in the same row is termed "**commutation**".

It is important to realize that the commutation process is a function of both circuit-dependent and switch-dependent parameters:

- ***Circuit-dependent parameters*** depend on circuit topology, and include components such as transformer leakage (inductor), commutation capacitor, auxiliary switching device, etc. For the 6-pulse bridge configuration, the most important circuit-dependent parameter for commutation is the finite transformer leakage (inductance); assuming typical values for this, an overlap angle $\mu < 60$ degrees is necessary, and more than two valves will conduct during the commutation period.
- ***Switch-dependent parameters*** include device turn-on and turn-off times, di/dt and dv/dt limitations, etc. The most significant switch-

dependent parameter is turn-off time, t_q. Since the switches are not perfect, their turn-off times are finite and impact on the commutation process. The switch turn-on time, however, is much smaller that the turn-off time of common power switches and does not impact on the commutation process in a significant manner. Depending on which type of switch is in use in the bridge, the type of commutation technique feasible is shown in Table 5-2.

Forced commutation techniques [1,2,3] may be applied either on the high-voltage (power) side of the converter by means of auxiliary components (i.e. thyristor, diode, inductor and/or capacitor), or alternatively on the low-voltage (controls side) of the converter using self-commutated devices.

Table 5-2: Commutation types

Type of device	Commutation	Initiated by
Conventional Thyristor	Line	AC Line voltage
	Circuit (or capacitor)	Capacitor voltage
GTO, IGBT or MCT	Self	Gate drive

For the commutation of the conventional thyristor converters, both circuit-dependent and switch-dependent parameters are critically important. Furthermore, for the conventional thyristor converter, it is possible to use either Line-Commutation (LC) or Circuit Commutation (CC) techniques.

For the commutation of the HVDC converter with, say GTO devices, the circuit-dependent parameters are now less crucial since the devices can be treated as perfect switches (within certain limits). For these newer devices, self-commutation techniques are employed.

5.2.1.1 Definition of Terms

In the past the terms "artificial commutation" and "forced commutation" have been used interchangeably. As a result of the new devices (i.e. GTOs, IGBTs), a new term called "self-commutation" was coined. A certain amount of confusion and misuse of these terms is apparent within the industry. These terms are defined below to clear misconceptions. These definitions are consistent with the IEEE Guide for Self-Commutated Converters ANSI/IEEE Std. 936-1987.

In this guide, conventional thyristors are called circuit-commutated devices, and GTOs, IGBTs and other such devices are called self-commutated devices. Artificial or Forced Commutation (FC) applies to both circuit-commutation using conventional thyristors, and self-commutation using GTOs and other devices. Although both circuit- and self-commutation techniques are examples of forced commutation techniques, the difference between circuit- and self-commutation is significant. These techniques are discussed below.

5.2.2 Line (or Natural) Commutation

This technique relies on the natural reversal of the sinusoidal ac line voltage across the valves of the converter. To initiate commutation, the firing pulse from the outgoing valve is removed and an alternate incoming valve in the same row is triggered to take up the dc current. During the commutation (overlap) period, the dc current is shared between the outgoing and incoming valves as a result of the leakage inductance of the transformer. Once current is transferred to the incoming valve, the reverse voltage across the outgoing valve is maintained for a time period t_{off} (equivalent to gamma angle); the outgoing valve must be reverse biased for a period greater than t_q, the turn-off time of the device. During this period a small reverse current is drawn from the device to deplete the charge carriers within the pn-junction of the device. The time difference between t_{off} and t_q is required to provide a margin of security for the device to achieve its voltage blocking capability. Typical valves of t_q and t_{off} are 350 μS and 700 μS respectively.

It is important to note that the voltage blocking capability of the device is a function of the (reverse-voltage * time period) product, and not the reverse-voltage alone. For example, low reverse-voltage for a long time period may not achieve successful blocking; similarly, high reverse-voltage for a short time period may fail to achieve voltage blocking capabilities of the valve due to high dv/dt stress.

With line commutation, because of the direct dependence of the firing angle alpha to the ac voltage, it is only feasible to delay the firing angle; it is not possible to advance the firing angle with reference to the ac system voltage. This means that alpha can vary only from 0 to 180 degrees; as is well known from converter theory, operation within these angles by a line commutated converter can only absorb reactive power from the ac system.

To advance fire the delay angle (i.e. from -180 to 0 degrees) it is necessary to have forced commutation, whereby the firing angle does not depend directly on the ac voltage; in this case, the converter can also supply reactive power to the ac system.

5.2.2.1 Limitations of Line Commutation

The fundamental limitation of a LC converter is its dependence on an adequate stiff ac voltage source for commutation purposes. Power systems are subject to disturbances, voltage regulation difficulties and harmonic pollution which cause commutation problems for such converters. As a result, LC converters have difficulties to feed into weak ac systems and may take prohibitively long times to recover from disturbances. Furthermore, the ability of the LC converter to control reactive power is limited.

These limitations can be overcome by the use of forced commutation employing either circuit- or self-commutation techniques.

5.2.3 Circuit Commutation

In case the ac line voltage is inadequate, distorted or sometimes even unavailable to achieve commutation of the conventional thyristor valves, circuit commutation may be used. An artificially generated voltage can be used to force commutate the valves. This artificially generated voltage is temporarily stored on a commutation capacitor until it is required to commutate the valve. This artificially generated voltage may be derived either from the following sources:

- The ac line voltage, whenever it is present,
- The dc line voltage, or
- An auxiliary voltage.

Commutation circuits deriving their energy from any one or multiple of these sources exist. The commutation circuit serves two distinct, but intertwinned roles:

- To provide the commutation voltage for the switching device, and
- To divert inductive load currents from the main switching device to another auxiliary switching device.

The first role is obvious and well understood by utility engineers. The current diversion role, however, is sometimes not fully appreciated by utility engineers, especially in relation to circuit-commutated devices. The significance of this current diversion role, however, becomes more apparent with the use of self-commutating devices.

The commutation capacitor could be either in series or in parallel with the main valve. Circuits of either type are feasible. It is noteworthy that all forced commutation circuits can be reduced to either one of these two types. General operating principles for these circuits are given below; specific examples of circuit which employ such techniques are provided in a later section.

5.2.4 Series Capacitor Circuit

A simplified equivalent circuit (Figure 5-1) show the commutation principle. When the main valve T1 is fired, load current i_L is established in the commutation capacitor C and one phase of the equivalent load. This equivalent load can be considered to be the resultant impedance of the load, ac filters and converter transformer. If no further circuit topology changes occur, the capacitor voltage v_c will eventually become greater than the dc line voltage V_d and the current i_L will be reduced to zero. Commutation is then completed.

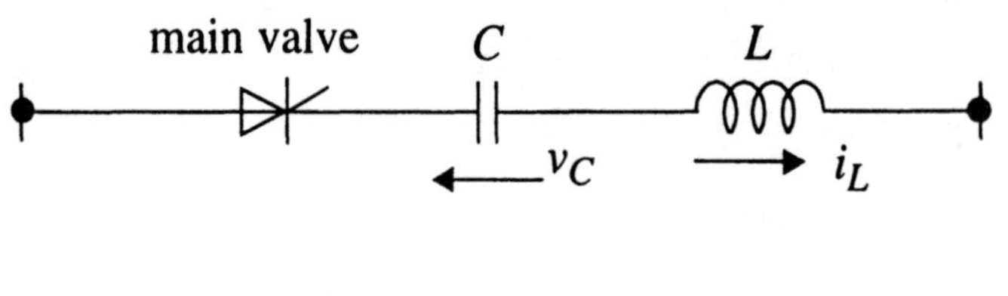

Figure 5-1: Commutation of main valve by a series capacitor

5.2.4.1 Parallel Capacitor Circuit

This circuit employs a commutating capacitor in parallel with the main valve (Figure 5-2). The circuit also employs an auxiliary valve CT1.

Load current into the equivalent load is established by firing main valve T1. To commutate the main valve, the auxiliary valve CT1 is fired. Assuming that the capacitor was pre-charged in the polarity indicated, the load current i_L will be diverted into the parallel path formed by CT1 and C; this will turn-off T1. At the same time, the capacitor will charge up in the opposite polarity.

This type of commutation circuit carries the load current only during the commutation period, unlike the series capacitor circuit above.

Minor variants of either of these two types of circuit exist for forced commutation purposes.

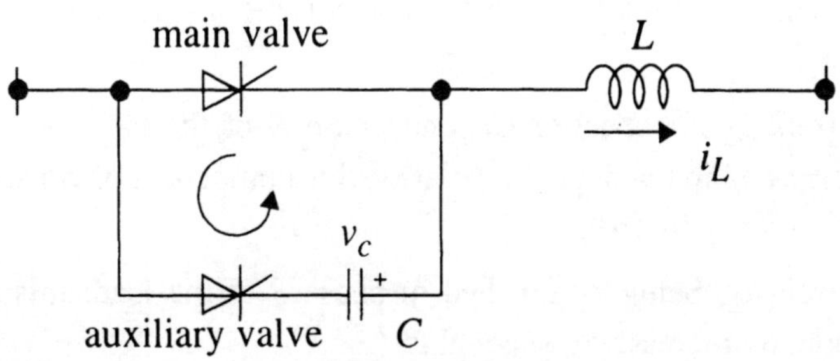

Figure 5-2: Commutation of main valve by parallel capacitor and auxiliary valve

5.2.5 Self-Commutation

Relatively newer devices (such as GTOs) are able to be turned off by application of a negative control pulse at their gates. Sometimes, just the removal of a gate bias voltage at the gate (or base) of the device may be enough to turn-off the device (such as a MOSFET transistor). This type of commutation is termed self-commutation and will be successful with power circuits having purely resistive (in phase) currents; however, since power circuits usually have inductive currents to be commutated, the transfer of current to another valve in the same row may not be successful unless additional circuits having diverters are utilized. The function of these diverters will be to temporarily divert the inductive load current to a capacitor, until the next incoming valve is able to pick up the current. The impact of such diverters has not yet been fully assessed by the industry.

It is practical to consider two versions of the converter for HVDC transmission: current source and voltage source.

5.2.5.1 Current Source Converter (CSC)

In HVDC transmission, this is the more traditional mode of power transmission.

The principle of a current source GTO converter is shown in Figure 5-3a. The dc line current is maintained constant by the use of a (large) smoothing reactor L_d. The main valve T1 is on and load current i_L is established in phase R of the load reactor L_r. The diverter capacitor C_{rs} is pre-charged in the polarity indicated. In order to self-commutate T1, the turn-off pulse to T1 is applied, and at the same time, the next phase main valve T3 is turned on. The main valve T1 is instantaneously turned off, subject to the dc line current being diverted to valve T3. The current in valve T3 (Figure 5-3b) is composed of:

- Current i_C in capacitor C_{rs} and phase R of the load; as the capacitor charges up in the opposite polarity, this component of current will gradually reduce to zero,
- Current i_{Lr} being established in phase S of the load; this current will gradually increase to be equal to I_d.

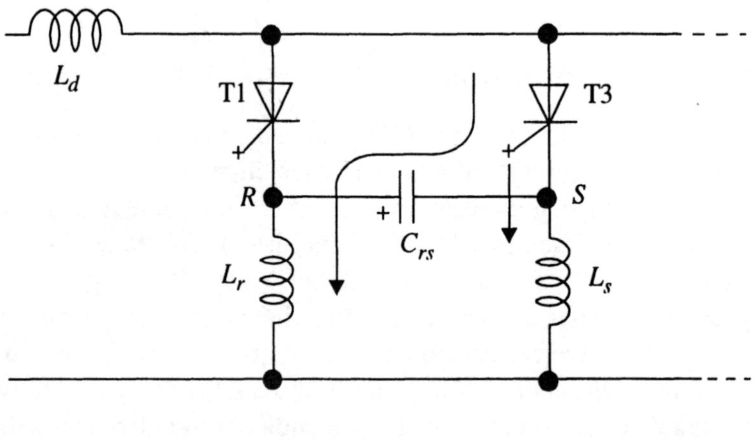

Figure 5-3a: Equivalent circuit for a current source inverter

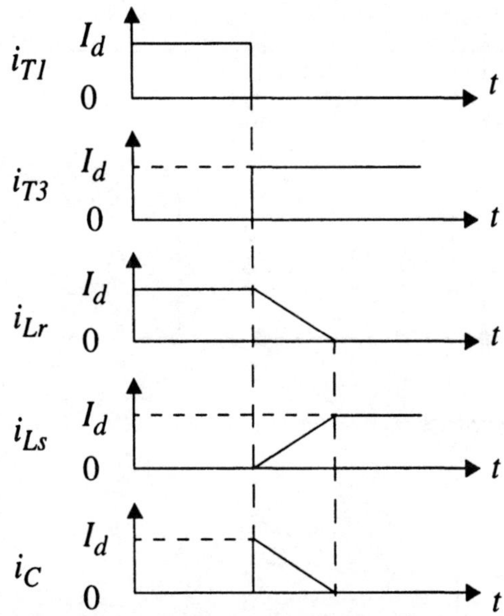

Figure 5-3b: Currents in equivalent circuit shown in Figure 5-3a

5.2.6 Voltage Source Converters (VSCs)

At the present time, this configuration can only be used for HVDC transmission below 250 MW rating due to switch rating limitations.

The principal of the voltage source converter is shown in Figure 5-4a. The circuit requires that the dc line voltage be maintained constant at the converter terminals. This is achieved by having a large capacitor C_{dc}, on the converter side of the dc smoothing reactor. In addition to C_{dc}, free-wheeling diodes D1, D3 etc., are required across the load phase *R* and *S* etc. to divert the load current during commutation of the main valves.

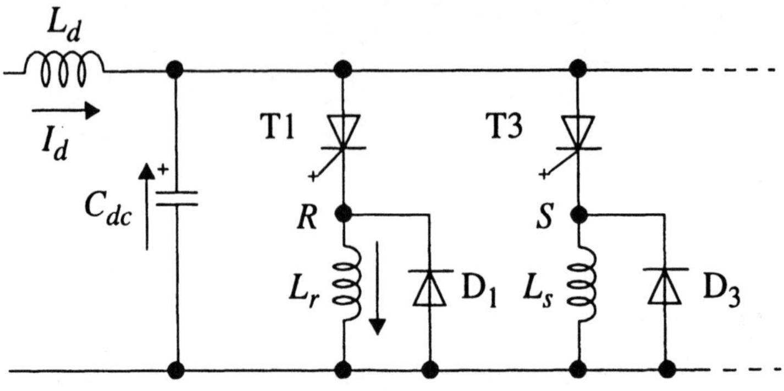

Figure 5-4a: Principle of the voltage source converter

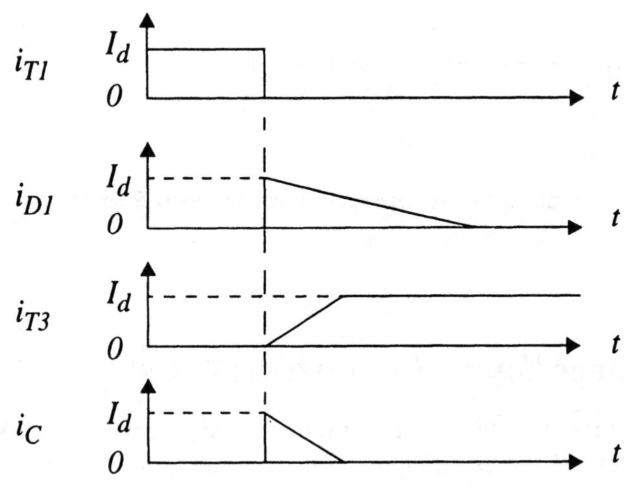

Figure 5-4b: Currents in equivalent circuit shown in Figure 5-4a

The main valve T1 is conducting current $I_d = i_R$ (Figure 5-4b) into phase R of the load. To commutate valve T1, a negative pulse is applied to its gate while the next valve T3 is fired. Valve T1 is turned off instantaneously, since the diode D1 is able to freewheel the load current in phase R; this current will decay at a rate depending on the resistance and inductance of the load. In the meantime, valve T3 is on and load current is being established in phase S of the load. The current difference between I_d and i_S is fed into the capacitor C_{dc}, until $i_S = I_d$.

5.2.6.1 Comparison of Current and Voltage Source Converters

A detailed comparison of current/voltage source converters has been provided in [4]. For this comparison, it is assumed that the same type of converter is available at both ends of dc system; this may not be the case in practice. Some of the major characteristics of systems with the two types of converters are listed in Table 5-3:

Table 5-3: Comparing characteristics of current and voltage source converters

Current Source Converter	Voltage Source Converter
Polarity of dc current is unidirectional.	Polarity of dc voltage is unidirectional.
Polarity of dc voltage changes with dc power flow.	Polarity of dc current changes with dc power flow.
DC smoothing reactor (large) maintains dc current constant.	DC capacitance (large) maintains dc voltage constant.
DC filter capacitance is used on line-side of smoothing reactor.	DC smoothing reactor is used on line side of dc filter capacitance.
AC-side diverter capacitors are required.	DC-side diode diverters are used in anti-parallel with main valves.
GTOs must be symmetrical reverse-blocking devices, or use diodes in series with the GTOs.	GTOs may be asymmetrical or symmetrical due to presence of diode diverters in anti-parallel with main valves.
The fault current contributed by the converter to a dc line fault can be limited by control action and minimized by the large dc smoothing reactor.	The fault current contributed by the converter to a dc line fault cannot be limited by control action since the diodes in the converter will feed into the fault. The fault current may only be interrupted by ac breakers, which is not very convenient for dc transmission purposes. Furthermore, the peak fault current may be excessive since the large dc filter capacitors will also discharge into the fault.

5.2.7 Regions Of Converter Operation

From converter theory, the relationship between dc voltage V_d and the firing angle alpha is

$$V_d = V_{do} \bullet Cos(alpha) - (3/(Pi)) \bullet X_c \bullet I_d) \qquad (5\text{-}1)$$

Converter operating regions are shown in Figure 5-5 as a polar plot of V_d versus alpha. These 4 regions are divided into four quadrants *Q1* to *Q4*. Theoretical and practical limits to these regions are defined in Table 5-4. Provided an adequate ac supply is available, line commutated (LC) converter operation is possible in quadrants *Q1* and *Q2*. In *Q1*, the converter operates as a rectifier consuming reactive power from the supply. The practical alpha-min limit (5 degrees) is required for the valves to have a forward-bias voltage before turning on. In *Q2*, the converter operates as an inverter again consuming reactive power from the supply. Generally, two limits apply in this region i.e. the alpha-min (105 degrees) and alpha-max (145 degrees) limits in inverter mode of operation; the alpha-min limit is imposed for operational reasons following recovery from dc line faults, etc., and the alpha-max limit is imposed to ensure commutation margin angle security. The practical operational region for a LC inverter is shown as region *X*. To operate beyond region *X* into quadrant *Q3* requires assistance from forced commutation.

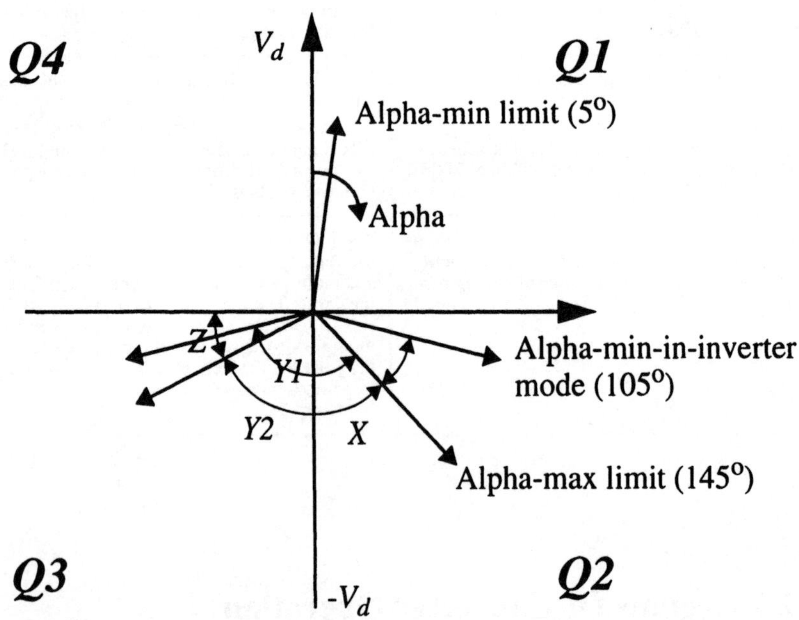

Figure 5-5: Regions of Converter operation

Theoretically, forced commutation permits operation in any quadrant. Depending on the commutation technique, practical limits for the FCCs also exist.

Table 5-4: Four quadrant operation

Quadrant	Converter operates as	Alpha limits (degrees) Theoretical	Practical
Q1	LC rectifier	1-90	5-90
Q2	LC rectifier	90-180	105-145
Q3	CC inverter	180-270	180-240
Q4	CC rectifier	270-360	-

5.2.7.1 With Circuit Commutated Devices

With a parallel-capacitor type circuit, a low dc voltage limit exists beyond which a commutation failure occurs due to insufficient charge on the capacitor (end of region *Y2*, alpha = 240 degrees). In quadrant *Q2* this limit merges into LC inverter operation region *X*. With a series-capacitor type circuit, the low voltage limit extends further (end of region *Y1*) depending on the size of the capacitor. Operation into region *Z* is possible with variant of the parallel-capacitor circuit which employs an auxiliary source for charging up the capacitor.

Operation into quadrant *Q4* will be as a CC rectifier supplying reactive power to the ac supply.

5.2.7.2 With Self-Commutated Devices

The operational limits due to low-voltage will not exist since the gate drive circuits are not functions of this voltage. Hence, FCCs with self commutated devices will have practically no limits in its operation in all four quadrants.

5.3 EXAMPLES OF FC CONVERTERS FOR HVDC TRANSMISSION

There are two categories of FC circuits i.e. circuit-commutated or self-commutated. These two categories of FCCs will be discussed next.

Although 6-pulse circuits are shown in the following examples for reasons of simplicity, it is understood that the more usual 12-pulse configuration will be used for practical schemes.

5.3.1 Circuit-Commutated Converters

5.3.1.1 Series Capacitor Circuits

Also known as the load-side commutation circuit, this circuit employs three capacitors [16,18,20,21,23,24] in series with the ac phase connections *R*, *S* and *T* to the 6-pulse bridge (Figure 5-6); no additional valves are required. This type of circuit may require an auxiliary supply for start-up purpose when feeding a dead load.

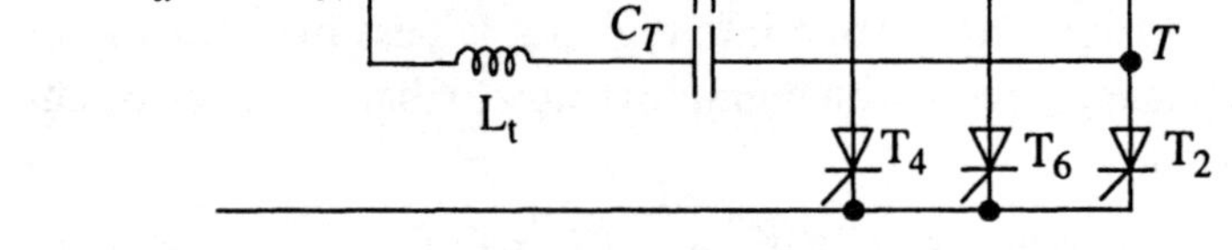

Figure 5-6: Series capacitor commutation circuit

A simplified equivalent circuit (Figure 5-7) shows the commutation principle. When valve T1 is fired, load current i_L is established in capacitor C_r and the T-phase of the equivalent load. This equivalent load can be considered to be the resultant of the load, ac filters and converters transformer. If no further circuit topology changes occur, the capacitor voltage V_{cr} will eventually become greater than V_d, and the current i_L will reduce to zero. Before this happens, valve T3 will be fired and load current in the S-phase will be established. Capacitor C_s had a charge of the polarity indicated from a previous cycle. Firing T3 will turn-off T1. This procedure is then repeated for the next commutation cycle using the required firing order.

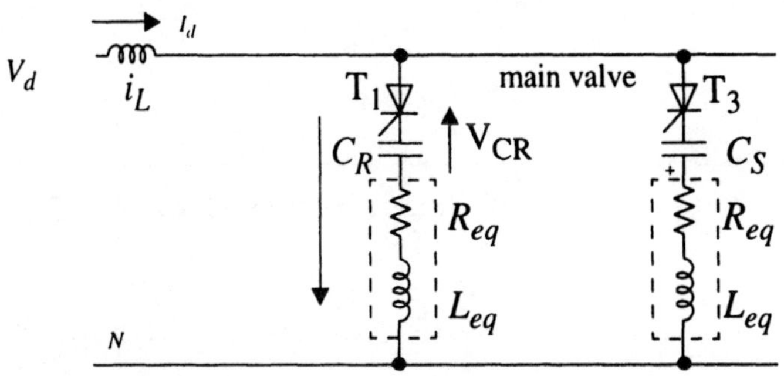

Figure 5-7: Equivalent circuit

5.3.1.2 Parallel Capacitor Circuits

Also known as the dc- or supply-side commutation circuit, this circuit employs the commutation capacitor in parallel with the main valves (Figure 5-8). The capacitor derives its energy directly from the dc line. The circuit employs one capacitor and two commutating CT1 and CT2 per 6-pulse bridge. This type of circuit does not rely on an auxiliary supply for start-up purposes when feeding a dead load.

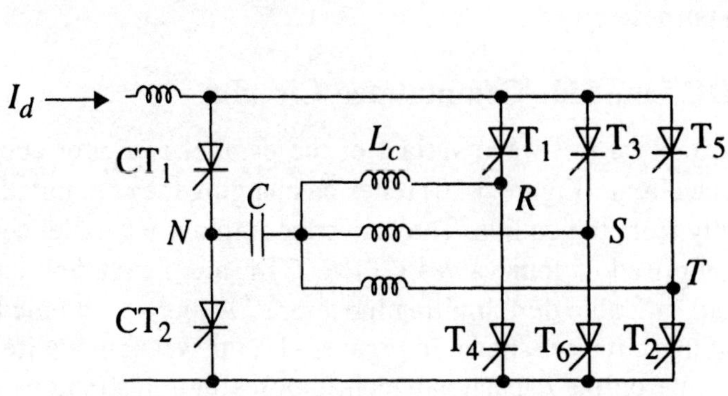

Figure 5-8: Parallel capacitor commutation circuit

The commutation principle is shown in Figure 5-9. Load current into the equivalent load is established by firing valve T1. To commutate T1, valve CT1 is fired. Assuming that the capacitor was pre-charged in the polarity indicated, the load current i_L will be diverted into the parallel path formed by CT1 and *C*; this will turn-off T1. At the same time, the capacitor will charge up in the opposite polarity, in readiness for the next commutation when valve CT2 will be fired.

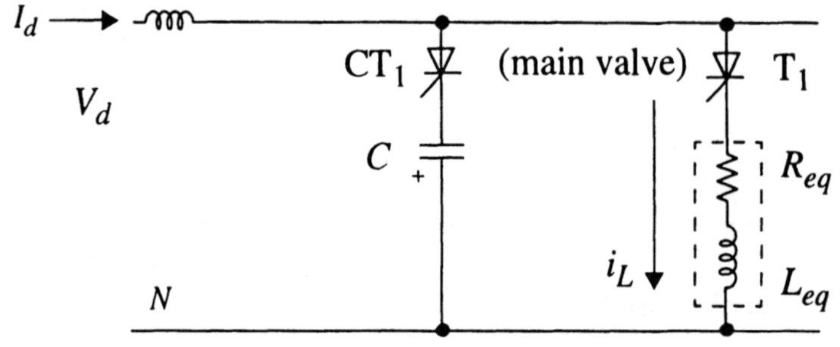

Figure 5-9: Equivalent circuit

This type of commutation circuit carries the load current only during the commutation period, unlike the series capacitor circuit above.

Minor variants of either of these two types of circuit exist for forced commutation purposes.

5.3.1.3 DC Line Side Commutated Circuits

This is actually a full-wave variant of the parallel capacitor commutation circuit. The circuit (Figure 5-10) relies on charging the commutation capacitor directly from the dc line. The converter employs a parallel commutator bridge comprised of four valves CT1 to CT4, a commutation capacitor *C* and a small saturable di/dt limiting inductor *L*. A more economic two-valve version of this circuit is shown in Figure 5-11; this version requires a charge reversal cycle on the capacitor which imposes time restrictions on circuit operation for high frequencies only. For power frequency operation, this version is quite feasible.

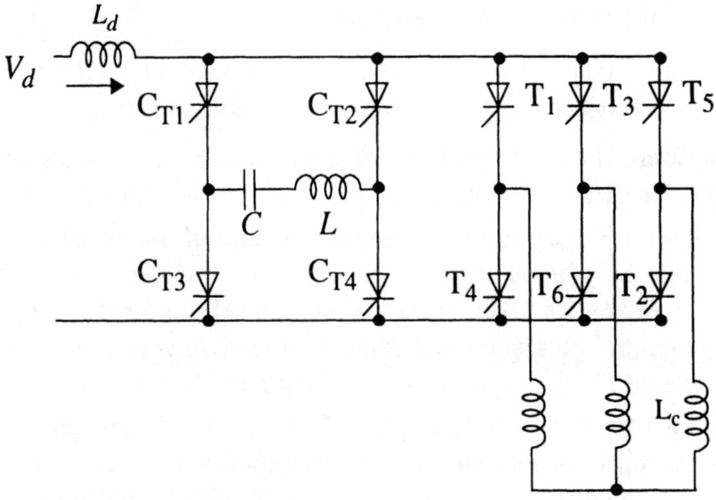

Figure 5-10: DC line-side commutated circuit with 4-valve version of commutator bridge

Figure 5-11: DC line-side commutated circuit with 2-valve version of commutator bridge

5.3.2 Self-Commutated Converters

5.3.2.1 Current Source Converter Circuit

In HVDC systems, this is the traditional mode of transmission. The dc current is maintained constant by use of a (large) smoothing reactor.

In Figure 5-12 are shown the position of three ac-side capacitor diverters C_{rs}, C_{rt} and C_{st} in delta-configuration (a star-configuration for these capacitors id also feasible). For example, consider the case of valves T1 and T2 conducting, and the commutation of valve T1 and transfer of current to valve T3. The valves are GTO based and can be self-commutated by the application of control pulses. For the commutation of valve T1 and transfer of current to valve T3, the capacitor C_{rs} will temporarily take over the current in the transformer inductance of phase *R*, until valve T3 and phase *T* is fully able to establish the current. The dc capacitor diverter will also assist in the transfer of the dc current from valve T1 to T3. The dimensions of the three ac side diverter capacitors are functions of the product L_c*I_d; clearly, a lower value of the inductance L_c will help in reducing the size of the capacitor. A typical value of the converter transformer leakage is 16%; this could be reduced to 10-12% to enable some reduction in the size of the diverter capacitors.

The design of the circuit should also consider the natural resonant frequency of the L_c*C_{rs} circuit which would interfere with the operation of the converter. The diverter capacitors would also tend to reduce the harmonics generated by the converter.

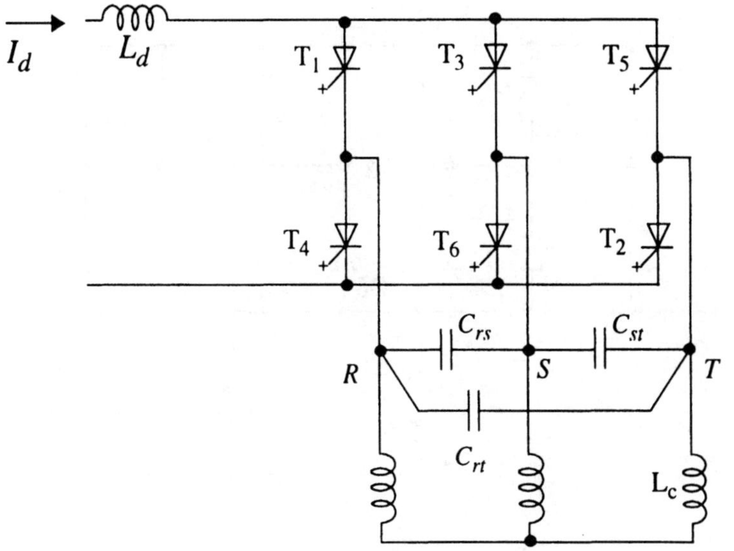

Figure 5-12: Current-source self-commutated converter

5.3.2.2 Voltage Source Converter Circuit

This configuration (Figure 5-13) is now used for HVDC transmission below 250 MW [5,6,19,23] using either GTOs or IGBTs as the main switches.

The circuit requires the dc voltage to be maintained constant at the converter. This is achieved by having a capacitor C_{dc} on the converter side of the smoothing reactor. In addition to C_{dc}, free-wheeling diodes D1 to D6 are required across each valve, to assist in the current diversion during commutation of the main valves.

The capacitor C_{dc} will also reduce the dc harmonics generated by the converter. Additionally, the capacitor C_{dc} will provide protection from line surges.

In order to take advantage of the fast switching capability of the GTOs, pulse width modulation (PWM) techniques can be utilized to reduce the low-order harmonics generated by the converter; this will reduce the ac filter cost.

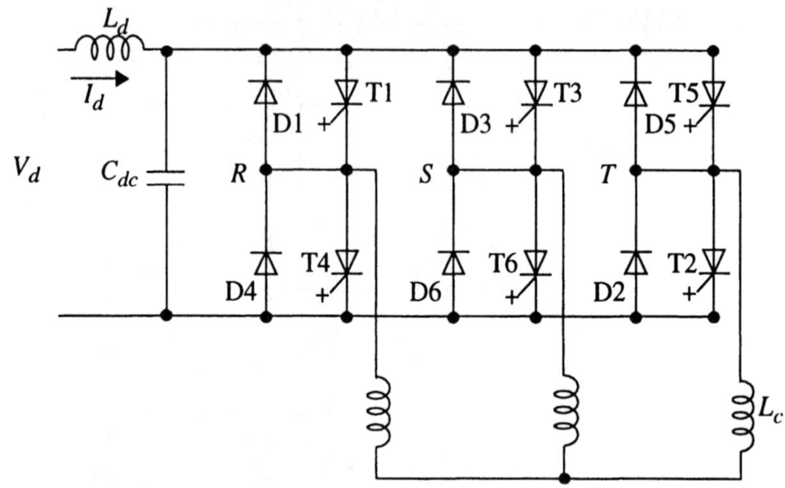

Figure 5-13: Voltage-source self-commutated converter

5.4 REFERENCES

[1]. F. Busemann, "Artificial Commutation of Static Converters", Electrical Research Association, England. Report. B/T 109, 1951.

[2]. J. Reeve, J.A. Baron and G.A. Hanley, "A Technical Assessment of Artificial Commutation of HVDC Converters with Series Capacitors", IEEE Trans. on Power Apparatus and Systems, Vol. PAS - 87, No. 10, October 1968, pp 1830 - 1840

[3]. V.K.Sood and J.Bowles, "Force-commutated HVDC Inverters". Canadian Electrical Association, Spring meeting 1979. (now out-of-print, call author for copy).

[4]. A.M. Gole and R.W. Menzies, "Analysis of certain aspects of forced commutated HVDC inverters" IEEE Trans. on Power Apparatus and Systems, Vol. PAS-100, No. 5, May 1981, pp 2258 -2262

[5]. R. Jotten and W. Michel, "Control with an Inverter Applying Forced Commutation". CIGRE SC 14 Meeting in Rio de Janeiro, August, 1981, Item 12.1.

[6]. K.S. Tam and R.H. Lasseter, "A Study of a Hybrid HVDC Converter". Int.

Conference on DC Power Transmission, Montreal, Quebec. June 4-8, 1984.

[7]. H.M. Turnali, R.W. Menzies and D.A. Woodford, "A Forced Commutated Inverter as a small Series Tap on a DC Link". International Conference on DC Power Transmission, Montreal, Quebec, June 4-8, 1984.

[8]. V.K. Sood. "A novel dc line-side force-commutated HVDC inverter for feeding remote loads". IEEE International Communications and Energy Conference, Montréal. 2-4 Oct. 1984. pp 86-89.

[9]. V. K. Sood. "Force-commutated HVDC Inverters". IEEE International Communications and Energy Conference, Montréal, 2 - 4 Oct. 1984. pp 90-93.

[10]. V.K. Sood, "A novel force-commutated thyristor inverter for a series tap in a HVDC line". Int. Conference on Computers, Systems and Signal Processing, Bangalore, India. 10-12 Dec. 1984.

[11]. V.K. Sood, "Analysis and Simulator Evaluation of a dc line-side Force-Commutated HVDC Inverter for feeding a remote load". IEEE Power Electronics Specialists Conference, Toulouse, France, 24 - 28 June 1985.

[12]. V.K. Sood, "An Introduction to Forced-Commutated HVDC Inverters". Canadian Electrical Association Spring Meeting, Montréal, 24 - 26 March 1985.

[13]. V.K. Sood, "Analysis and Simulator Evaluation of a small Force-Commutated Series Inverter Tap in a HVDC Line". IEE Fourth International Conference on AC and DC Power Transmission, London. 23 - 26 Sept. 1985.

[14]. W.McMurray and H.Mehta, "Feasibility of Gate Turn-Off Thyristors in a High Voltage Direct Current Transmission System," EL-5332, EPRI Research Report project 2443-5, Final Report August 1987.

[15]. V.K. Sood, Position paper for Canadian Electrical Association on "Artificially Commutated HVDC Inverters", March 1989, Contract No. ST-174B.

[16]. Tomas Jonsson, Per-Erik Bjorklund, "Capacitor Commutated Converters for HVDC", Stockholm Power Tech, June 1995, Proceedings: Power Electronics, Int. Symposium on Electric Power Engineering, Royal Institute of Technology and IEEE, pp 44-51

[17]. T.Holmgren, G.Asplund, S.Valdemarsson, P.Hidman, U.Jonsson, O.Loof, "A test installation of a self-tuned ac filter in the Konti-Scan 2 HVDC link", 1995 Stockholm Power Tech, Int. Symposium on Electric Power Engineering, Royal Institute of Technology and IEEE,

[18]. K. Sadek, M. Pereira, D.P. Brandt, A.M. Gole and A. Daneshpooy, "Capacitor Commutated Converter Circuit Configurations for DC Transmission", IEEE Transactions on Power Delivery, Vol. 13, October 1998, pp 1257 - 1264.

[19]. H.Jiang and A.Ekstrom, "Multi-terminal HVDC systems in urban areas of

large cities", IEEE Trans. on Power Delivery, Vol. 13, No.4, October 1998. pp 1278-1284.

[20]. M. Meisingset, A.M. Gole, R. Burton, O. Eide, R. Fredheim, "Impact of Capacitor Commutated Converters in AC Systems with Multiple DC Infeed", Conference Proceedings, the 13th PSCC (Power Systems Computation Conference), Vol. 1, pp. 516 - 522, Trondheim, Norway, June/July 1999.

[21]. Tomas Jonsson, Per Holmberg, Thomas Tulkiewicz, "Evaluation of Classical, CCC and TCSC Converter Schemes for Long Cable Projects", presented at EPE 99, Lausanne Switzerland, September 1999

[22]. Tsuyoshi Funaki, Kenji Matsuura, "Predictive Firing Angle Calculation for Constant Effective Margin Angle Control of CCC-HVdc", IEEE Trans. on Power Delivery, July 2000, Vol.15, No.3, pp 1087-1093.

[23]. F.Schettler, H.Huang and N.Christl, "HVDC Transmission systems using Voltage Source Converters - Design and Applications", IEEE Summer Power Meeting, July 2000, Vol.2, pp 715-720.

[24]. A.M. Gole, M. Meisingset, "Capacitor Commutated HVDC Converters for long cable HVDC transmission" Power Engineering Journal, June 2002, Vol.16, No.3, pp 129-134

Chapter 6

Capacitor Commutated Converters for HVDC Systems

6.1 CAPACITOR COMMUTATED CONVERTERS

This "new" HVDC converter configuration has been discussed about for more than 50 years! One of the earliest studies of this configuration was first carried out in the 1950s [1]. A more detailed investigation into the operation of this circuit was reported in [2]. In the late 1970s and early 1980s, studies with the parallel capacitor version were reported [3,4,5,6]. However, due to practical limitations with the valve ratings, these studies did not lead to any actual installations.

After a brief period of activity [3,4] while some attempt was made to consider the system implications of such a converter (Figure 6-1), references to this converter configuration disappeared from the research field for a while since it offered a more costly and "difficult" operational alternative to the line commutated converter (LCC). However, in the early 1990s, the CCC configuration [5] was resurrected again due to a number of reasons:

- Problem of voltage ratings of valves became less of a constraint financially (due to increasing ratings and decreasing valve costs),
- Management of reactive power and high performance harmonic filtering could be dealt with independently due to the development of the continuously tunable ac filter and active filters, and
- Utility demands for operation with increasingly weaker ac systems has meant that commutation with LCCs has become much more unpalatable.

The CCC is characterized by having capacitors inserted in between the converter transformer and the converter valves. Thus, this capacitor is in series with the leakage impedance of the transformer and the main valves. This has a two-fold effect:

- The capacitor provides a forced commutation facility to the main valves (as explained earlier in another chapter), and
- The capacitor compensates for the leakage inductance (or reactive power demand) of the converter transformer.

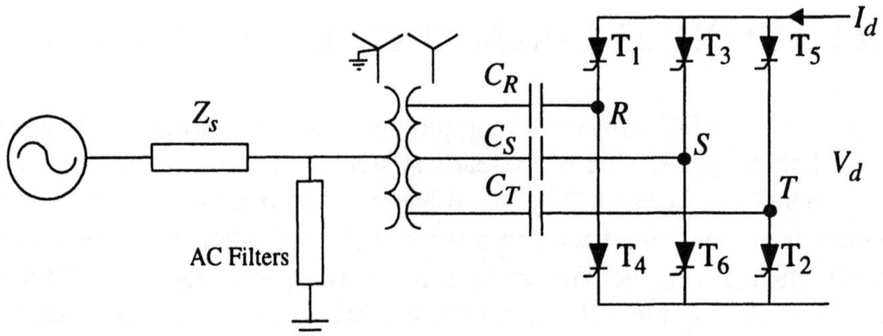

Figure 6-1: CCC circuit diagram

Sizing of the commutation capacitor, therefore, becomes a very important criteria as it impacts on the above two effects. A too-small capacitor will cause a large overvoltage across the capacitor (and valves), and not compensate sufficiently for the leakage inductance to result in a lagging current drawn from the ac bus. A too-large capacitor will result in low overvoltages and over-compensate for the demanded reactive power and might even draw a leading current from the ac system. However, a too-large capacitor also has a cost penalty associated with it. Common design practise suggests that an economical capacitor size would be to cause, say, a 10% overvoltage across the capacitor (and valves) as well as compensate for the reactive power demand to present a unity power factor to the ac bus. This design criteria maintains the cost of the valves at a reasonable level.

The capacitor voltage v_C is directly proportional to the dc current I_d, the (fixed) time t of conduction of the valve and inversely proportional to the size of the capacitor C i.e. $v_C = (I_d * t)/C$. Since v_C increases with I_d, this results in an increase of the dc voltage. This is in direct contrast to the case of a conventional LCC working at minimum extinction angle control where the dc voltage decreases with increase in I_d. This feature, therefore, results in a positive inverter impedance characteristic for the CCC providing improved dynamic stability (Figure 6-2). Therefore, the CCC can operate with very weak ac networks, and it can also tolerate a sudden drop of 15-20% in the ac bus voltage without suffering a commutation failure.

The capacitors are protected against overvoltages by parallel ZnO varistors.

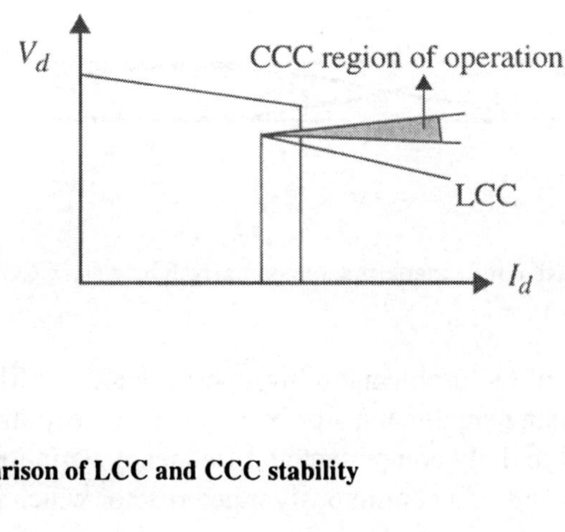

Figure 6-2: Comparison of LCC and CCC stability

6.1.1 Reactive Power Management

A comparison between the reactive power management of a LCC and a CCC are shown in Figure 6-3. The LCC serves the same need by means of switchable shunt capacitor banks. Since the CCC provides reactive power compensation Q proportional to the load current I_d of the converter, the need for switchable banks is eliminated. Consequently, ac filters are needed only for harmonic filtering; so their design can be optimized for this duty alone. Typically, only about 13% of the reactive power is required to be supplied by this minimal filter.

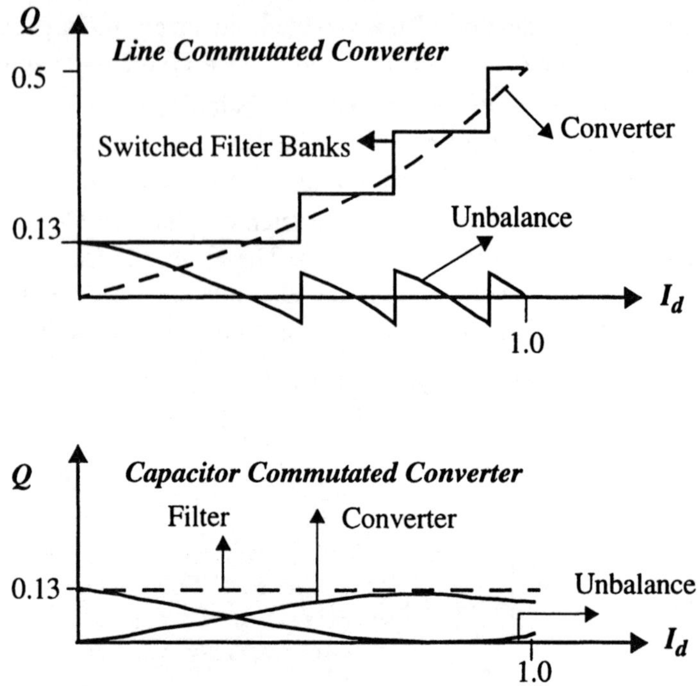

Figure 6-3: Reactive power management comparison between CCC and LCC [15]

In the past, one of the problems of high-performance ac filters (which were sharply-tuned using small capacitor banks) was to keep them in tune while being subjected to daily component and frequency variations. This problem was resolved by use of a continuously tuned reactor which is controlled by a dc current fed into a control winding mounted perpendicular to the main winding, enabling continuous adjustment of its inductance and thus continuously tuning the filter branch.

6.1.2 Thyristor Valve Modules

To optimize the cost of the thyristor valves (and the dc side equipment), the nominal dc current is optimized to be as close to the limit of the thyristor current rating at the permitted valve cooling limit. This permits the rated dc voltage to be kept low to achieve the rated dc power. A low dc voltage rating is beneficial for a compact modular valve housing as air clearances can be kept small.

The thyristor valves are air-insulated at atmospheric pressure and installed in modular valve housings. Each valve module contains two single valves, i.e. three modules for a 6-pulse converter. The thyristor valves are suspended from the ceiling and easily accessible for maintenance purposes. The surge arresters across the valves are also included in the housings.

6.2 CONTROLLED SERIES CAPACITOR CONVERTER (CSCC)

A modification of the CCC configuration has been proposed [6] where the series capacitor is re-located to be beyond the filter bus and in series with the source impedance (Figure 6-4a). This arrangement can be considered as an amalgamation of the CCC and LCC configurations, and the results obtained from steady state and dynamic performances confirm this. The advantage of this configuration is that the converter is a standard LCC.

In addition, the series capacitors can be controlled similar to a thyristor controlled series compensation (TCSC) scheme (Figure 6-4b). This variant is called the "Controlled Series Capacitor Converter" (CSCC).

6.3 COMPARISON OF CCC AND CSCC

A detailed steady state and dynamic performance comparison of the two systems was carried out in [6].

6.3.1 Steady State Performance

For the purposes of comparison, the two systems shown in Figures 6-1 and 6-4 are connected to a 300 kV ac bus and provide equal amounts of reactive power (116 MVAr referred to the 300 kV bus) from the ac filters and with the same value (22°) for the steady-state extinction angle of the valves at rated conditions. The dc systems in either case are rated at 500 kV, 1.6 kA.

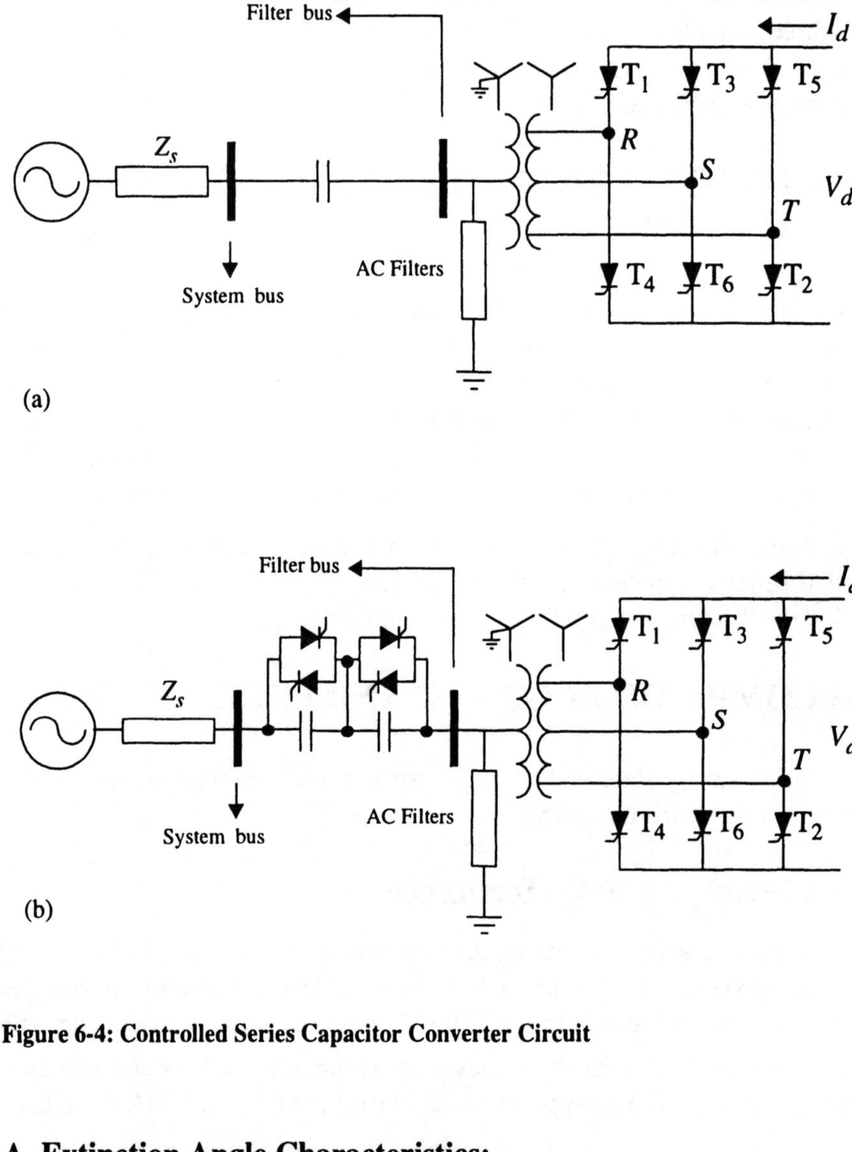

Figure 6-4: Controlled Series Capacitor Converter Circuit

A. Extinction Angle Characteristics:

One advantage of these topologies is that the series capacitors assist in the commutation process. Thus the apparent extinction angle viewed from the 300 kV ac bus bar can approach very small, or even negative values depending on the size of the selected series capacitor. The apparent extinction angle γ_{app} is the electrical angle corresponding to the time at which the valve turns off to the positive zero crossing of the corresponding apparent commutation (line-line) voltage on the ac bus bar. The actual extinction angle γ_{act} is larger because the real commutation voltage is the sum of the line-line ac bus bar voltage and the series capacitor voltages. The selected operating point has a value for $\gamma_{app} = 2^{o}$, which corresponds to an actual γ of 22°. The small value of γ_{app} results in an improved power factor and diminishes the requirement for shunt reactive power compensation.

Because the voltage on the series capacitor actually increases with dc current, the natural tendency for the extinction angle on an increase in dc current is to increase. This is the converse of the situation for the conventional converter in which an increase in dc current decreases γ, thereby bringing the converter closer to its commutation failure limit. This characteristic of the CCC or CSCC options is very favorable particularly for long cables. In these cases, a sudden lowering of inverter ac voltage, say due to a remote ac fault results in a sudden increase in dc current. The current controller on the rectifier has a negligible effect on this over-current which is primarily due to a discharge of the cable capacitance. The probability of commutation failure is reduced due to the natural tendency for γ to increase with increasing dc current.

Theoretical relationships found in [1,4] of the real extinction angle as a function of dc current for the CCC and CSCC options are plotted (Figure 6-5) assuming a control mode of constant $\gamma_{app} = 2^{o}$. It is noted that the extinction angle increases with dc current and that either option gives essentially the same result.

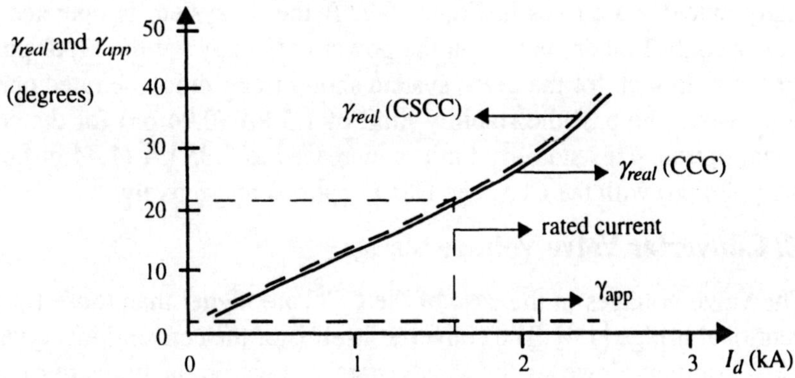

Figure 6-5: Comparison of actual and apparent extinction angles [6]

B. Maximum Available Power:

Assuming the inverter to be in Constant Apparent Extinction Angle (CAEA) control (γ_{app}= 2°), a theoretically calculated plot of dc voltage versus dc current (Figure 6-6) shows that the dc voltage for both the CCC and CSCC options has a much smaller slope as compared with the conventional LCC case.

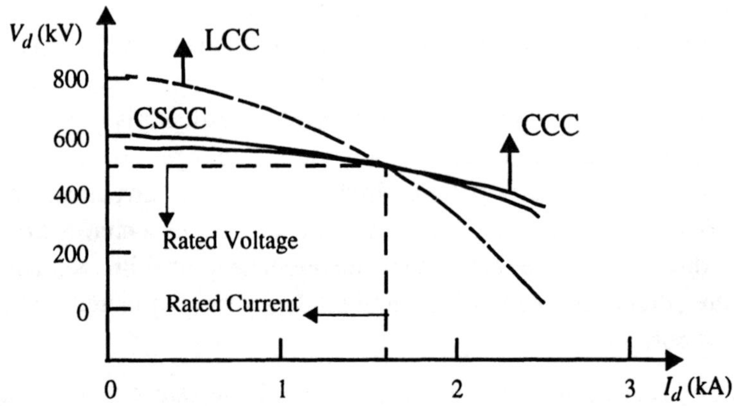

Figure 6-6: DC voltage versus dc current for CEA control mode [6]

This gives a much larger value for Maximum Available Power (MAP) as compared to a LCC (operated at a typical γ =18°), as seen from the theoretically calculated curves in Figure 6-7. If the dc system is operated in the power-control mode, points on the power curve beyond the MAP point are unstable. In fact, for the given system short circuit ratio, the rated operating point would be past the stability limit of 1.5 kA (0.94 pu) for the conventional option. This stability limit is increased to 2.35 kA (1.44 pu) and 2.2 kA (1.34 pu) with the CCC and CSCC options respectively.

C. Converter Valve Voltage Stress:

The valve voltages in the case of the CCC are higher than those for a conventional bridge [1,4]. The converter itself is of the conventional type in the CSCC option. However the steady state voltage on the converter (ac filter bus) is higher (327 kV) than the rated voltage (300 kV) of the system bus. It turns out that considering the transformer turns ratio, the final valve voltages with the CSCC option is the same as in the CCC option at rated conditions. However, for operation with higher ac bus voltages, when the slightly different tap changing regimes for the two options are taken into account, the CCC option is seen to require a somewhat larger valve voltage rating.

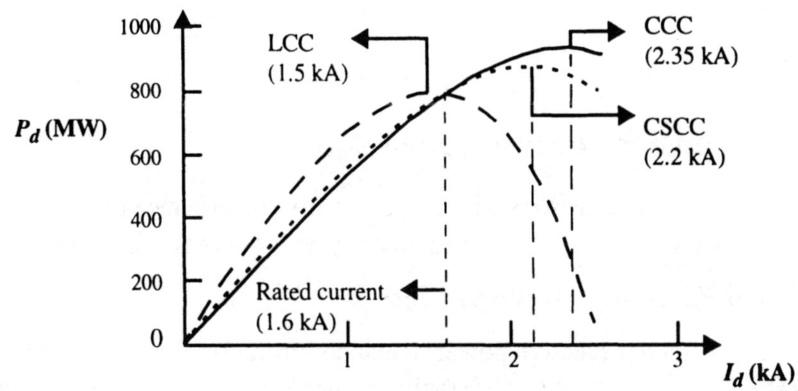

Figure 6-7: Maximum Available Power versus I_d for CEA control mode [6]

D. Harmonics and Filtering:

Since the ac filters are required only for harmonic elimination and not for reactive power support, the MVAr rating of the filter is reduced to very small values, which results in a very narrow passband. To keep the filter in tune for frequency or component variations, one option is to have a continuously tuned filter [4]; another option is to use active filters [11]. Unlike the conventional case, neither option requires filter bank switching for variations in the load over the full range of operation which simplifies the switchyard design. In this study, the filter MVAr rating is selected to be about 15% of the rated dc power.

On account of the smaller overlap angles that results because of the additional commutation voltage provided by the capacitors, the dominant current harmonics (11th and 13th generated in both options are typically higher than that for the conventional dc installations. Table 6-1 shows the characteristic converter current harmonic magnitude for the CCC and CSCC options at the rated operating point. The CSCC appears to have a smaller harmonic content when compared with the CCC option.

Table 6-1: AC current harmonics [6]

Configuration	Magnitude (A)			
Harmonic #	11	13	23	25
CCC	182.8	119.6	31.6	27.7
CSCC	166.3	111.0	9.4	9.0
LCC	140.9	102.0	18.0	14.2

6.3.2 Transient Performance

In this section, comparisons with respect to the performance under transient conditions such as load rejections and ac system faults are discussed.

A. Load Rejection Over-voltages:

Due to the smaller reactive power demanded from the filter bus, both CCC and CSCC options have considerably smaller load rejection over-voltage as compared with a LCC. This is particularly so for a weak ac system in which the equivalent impedance of the ac system is large [10]. For the test system, studies indicate that the magnitude of the load rejection over-voltage for the CCC and CSCC are approximately the same (1.17 pu), but are smaller than the over-voltage for the LCC case (1.4 pu).

B. Three Phase AC Bus Fault:

Typical faults were applied to the two options -- the LCC and CCC -- in order to evaluate the recovery performance of the two systems [12]. No special controls were modeled in this exercise. Figure 6-8 shows (a) dc voltage, (b) dc current and current order, and (c) inverter ac bus voltages for the LCC and CCC options respectively.

The fault is applied at 0.05 ms into the run and has a 5 cycle duration. After an initial over-current, the dc current is brought to zero due to the VDCL action taken at the rectifier end. The nature of the recovery, i.e. peak over-current, over-voltage and recovery is very different in both options.

The LCC option exhibits a slightly under-damped transient during the fault, but has a smooth recovery. The CCC option is more damped during the fault period, but has a small oscillation during the recovery. Both options restore the power to 90% within 150 ms after fault clearing. It is noteworthy that the CSCC option (not shown here) has a similar behavior as the CCC option.

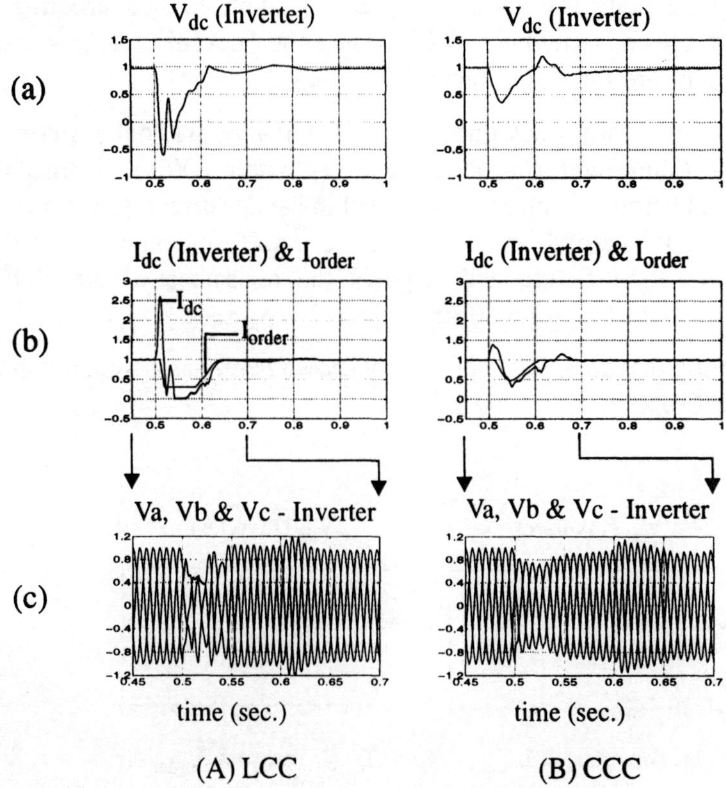

Figure 6-8: Three phase ac bus fault - LCC and CCC options [12]

C. Single Phase Remote AC Fault:

One of the shortcomings of the LCC when used in a long dc cable system is that ac side voltage reductions can cause the dc cable to discharge, transiently increasing the dc current in the inverter. The natural instantaneous effect of a current increase is a loss of commutation margin and hence an increased probability of commutation failure. The CCC and CSCC options have the opposite tendency, in that the instantaneous response to a current increase is an increase in the extinction angle.

In this case, the performances of the LCC and CCC options are compared with the application of a remote ac fault. The dc system is modeled as a long cable, which would discharge into the inverter because of the resulting reduced ac voltage. The inherent characteristic of the converter of increasing γ for increasing dc current allows the CCC option to operate successfully through the fault. The single phase fault is simulated by connecting an impedance to the converter bus so as to reduce the Thevenin source voltage by 20%. Results are shown in Figure 6-9.

The CCC option easily rides through the disturbance without suffering a commutation failure with full power recovery within 200 ms. During the fault, a second harmonic current is observed in the dc current; this is a characteristic of an unbalanced fault of this type. The LCC option, however, suffers a commutation failure and requires the assistance of the VDCL protection to reduce the current order.

It is noteworthy that the CSCC option (not shown here) has a similar behavior as the CCC option.

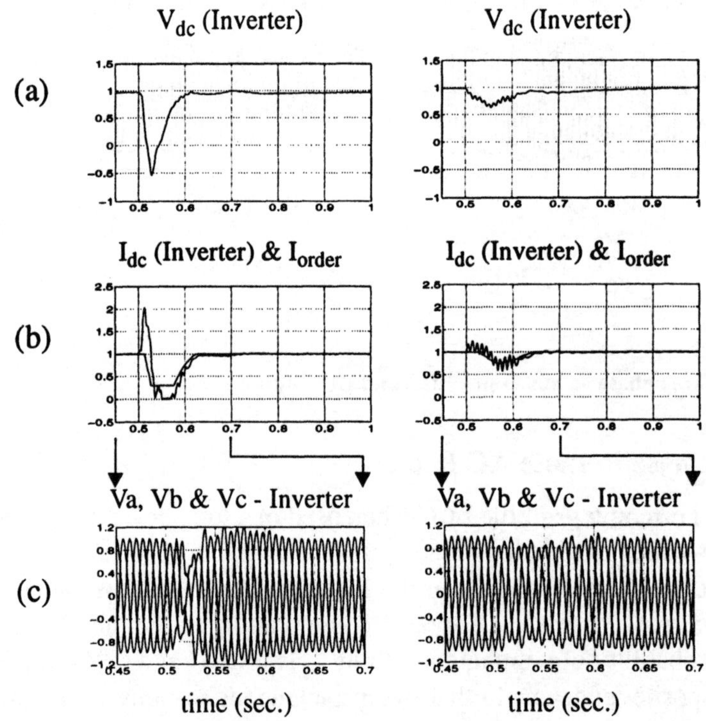

Figure 6-9: Single phase remote ac fault - LCC & CCC options [12]

D. Valve Short Circuit Over-current:

For rectifier operation with the CCC option, the series capacitor significantly reduces the valve short circuit current. In the CSCC option, the series capacitor is not directly between the filter bus and the converter valves, and consequently the short circuit current is larger than in the CCC option. Nevertheless, as shown in Figure 6-10, the magnitude and duration of this current is still smaller than that of the valve short circuit current in a conventional converter.

The performance of the CCC and CSCC options is very similar for steady state as well as transient operation. The maximum valve voltage and the ac current harmonics for the CSCC option are lower than the CCC option. On the other hand, the CCC option in rectifier operation exhibits a smaller valve short circuit current.

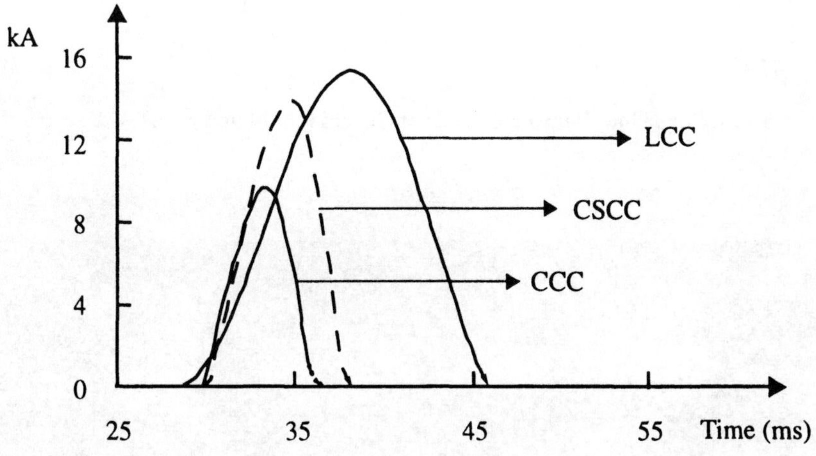

Figure 6-10: Valve short circuit currents [6]

6.4 GARABI INTERCONNECTION BETWEEN ARGENTINA - BRAZIL

The first commercial installation of the CCC type was at the 1100 MW Garabi back to back (BB) interconnection between the 500 kV systems of Argentina and Brazil [13]. Since the Argentinian system is at 50 Hz while the Brazilian system at 60 Hz, a BB frequency converter installation was necessary. Furthermore, since the short circuit levels at the converters were low, a CCC option was selected to provide the enhanced stability for the ac systems due to the forced commutated converters. A second interconnection of another 1100 MW is under construction. A single line diagram and an aerial view of the installation is shown in Figure 6-11.

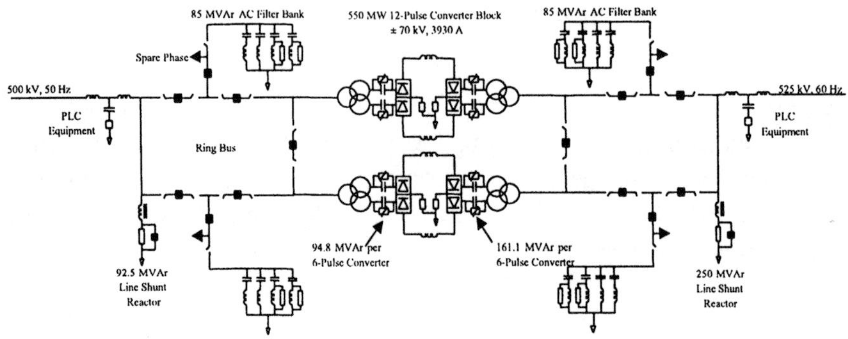

Figure 6-11a: Single line diagram of CCC at Garabi [13, 14 and ABB]

Figure 6-11b: Aerial view of CCC at Garabi [13, 14 and ABB]

The principal reasons for opting for the CCC option were:

- Switchable shunt filter banks were not required and replaced by the series capacitor bank to compensate for the reactive power,
- The CCC alternative provided a dynamically more stable operation with the weak ac systems,
- Since the series capacitor's impedance is typically several times greater than the transformer's leakage impedance, it reduced the valve currents during dc side short circuits allowing optimization of the transformer and valves, and
- The power factor, seen from the ac bus, could be kept close to unity, or even become positive during certain operating ranges.

Since this installation provides a ground breaking departure from conventional BB stations, some of its unique and innovative aspects are discussed next.

6.4.1 Valve Stresses

Due to the higher impedance of the capacitor as compared to the leakage impedance of the transformer, the short circuit currents are considerably reduced (by a factor of between 2-3 times) in the CCC concept as compared to the LCC. This allows the transformer and valve current ratings to be optimized for lower cost.

On the other hand, the voltage across the commutation capacitor results in higher peak voltages across the valves. The capacitors themselves need to be protected from overvoltages by means of Zn0 varistors across them.

Since the series capacitor reduces the overlap angle due to compensation of the leakage inductance of the transformer, switching voltage stresses and losses are reduced.

6.4.2 AC Switchyard

The reactive power management of the CCC is far superior when compared to a LCC. The elimination of the switches for reactive power compensation equipment simplifies the design, layout and space requirements of the ac switchyard. Since no ac breakers or disconnects, apart from the energizing purposes (Figure 6-12) are needed, system reliability is improved.

Figure 6-12: Special T-Circuit breakers used at Garabi Converter Station [ABB]

6.4.3 AC Filters

Since the reactive power requirements of the station are now mostly met by the series capacitors, the filtering requirements can be uniquely met by the ac filters. Typically, the ac filters of a CCC provide less than 15% of total converter reactive power demand as compared to about 55% for a LCC. This means that the ac filters need to be high performance units to optimize the costs. Therefore, electronically controlled filters which are sharply tuned (Q factor ~ 150, and therefore lower losses) can be used for the 11th and 13th characteristic harmonics, which are the dominant harmonics for a 12-pulse converter. By contrast, conventional band pass filters have to be equipped with damping resistors to give a broad characteristic to allow them to perform within the frequency and component value variations (Figure 6-13).

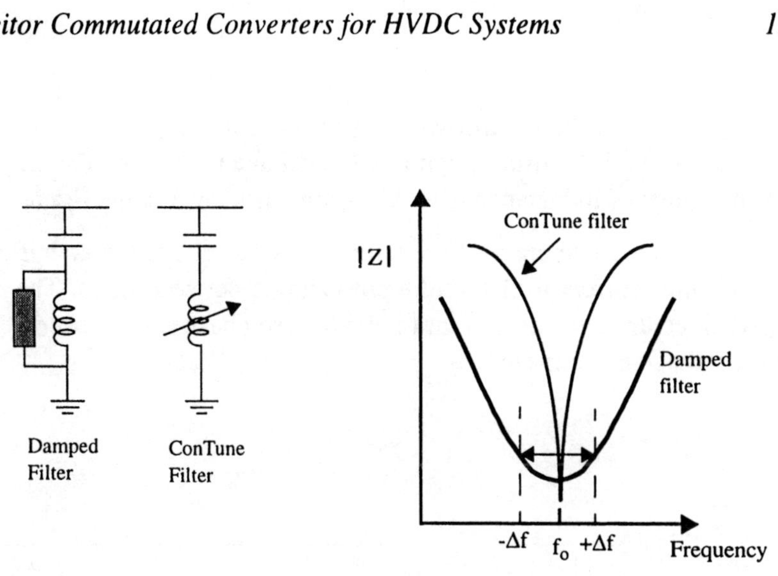

Figure 6-13: Comparison of conventional Damped filter and ConTune filter

Since the CCC acts like a static compensator, giving smooth continuous control of voltage and power flow, the minimum size of the ConTune filters help to keep load rejection overvoltages within limits. Therefore, switching banks are not needed.

The adjustable reactors of the filters are Continuously Tuned (hence the trade name "ConTune") by a dc current fed into a control winding mounted perpendicular to the main winding (Figure 6-14) and coil arrangement.

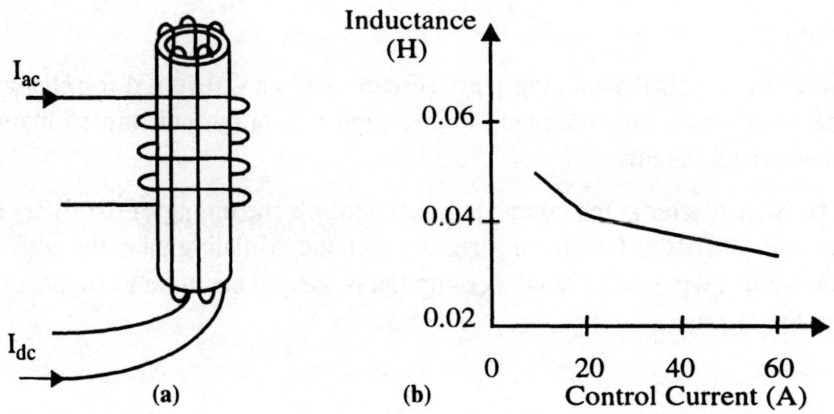

Figure 6-14: ConTune filter (a) structure and (b) linearity [15]

The dc current in the control winding influences the magnetic flux in the iron core. A stable control design is achieved due to the high linearity of the rate of change of inductance with dc current, as shown in the figure.

The control loop (Figure 6-15) requires measurement of the ac bus voltage and the filter current to derive the phase angle between them. The control loop then controls the dc current to obtain zero phase shift between the harmonic voltage and current.

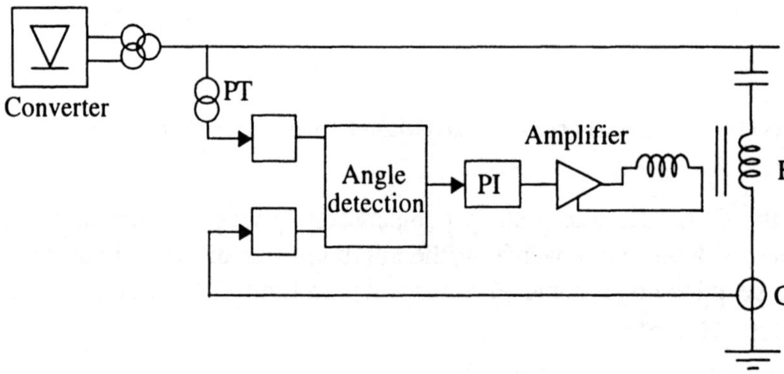

Figure 6-15: Control circuit of ConTune filter

Since no physically moving parts (Figure 6-16) are required for changing the inductance, the equipment enjoys high reliability and limited maintenance requirements.

The filter reactor is the component that controls the tuning of the filter. The reactor consists of four main parts: the coil, the insulating tube, the core and the control winding. A small cooling fan is used to maintain a compact size for the reactor.

Figure 6-16: The ConTune static ac filter [ABB]

6.4.4 Thyristor Valves Modules

To minimize the cost of the valves and dc equipment, the dc current is kept as high as possible within the thermal capabilities of the thyristor and cooling system. This means that for a given dc power, the dc voltage can be maintained low permitting a small air clearance requirement resulting in a compact modular design for the valve housing.

The thyristor valves are air insulated at atmospheric pressure and installed in modular valve housings. Each valve module contains either two or more single valves, which implies up to 6 valve modules per 12 pulse converter. The thyristor valves are suspended from the ceiling and are easily accessible for maintenance purposes. The housing also contains the surge arrestors connected across the valves. (Figure 6-17).

Figure 6-17: Thyristor Valves enclosed in containers at Garabi [ABB]

6.4.5 Modular Design Benefits

The modular design of the BB station results in saving of time and cost due to following factors:

- No need for a valve hall, control and service buildings. Therefore, the design and cost of the civil structure is kept to a minimum,
- Installation/commissioning time can be reduced considerably since most of the assembly/testing work is performed at the factory,
- A standardized design of the equipment reduces the number of spare parts required which facilitates storage and handling of spares, and
- The modular converter station is compact enough to be fitted into an existing right of way of a typically 400 kV ac line.

6.5 CLOSING REMARKS

The cost of a CCC is presently more than a comparably rated LCC by a factor of about 25%. This cost differential is likely to decrease with time as the cost of the valves will continue to decrease. However, the CCC offers features that the LCC cannot. For this reason, the CCC offers an attractive solution for inter-connecting weak ac systems, and for dc systems with a long dc cable.

6.6 ACKNOWLEDGEMENT

The considerable support of Mr. N. Ottosson of ABB and Mr. P. Lips of Siemens in providing data and graphics used in this chapter is gratefully acknowledged.

6.7 REFERENCES

[1]. F. Busemann, "Artificial Commutation of Static Converters", Electrical Association, England. Report. B/T 109, 1951.

[2]. J.Reeve, J.A. Baron and G.A. Hanley, "A Technical Assessment of Artificial Commutation of HVDC Converters with Series Capacitors", IEEE Trans. on Power Apparatus and Systems, Vol. PAS - 87, no. 10, Oct. 1968, pp 1830 - 1840.

[3]. A.M. Gole and R.W. Menzies, "Analysis of certain aspects of forced commutated HVDC inverters" IEEE Trans. on Power Apparatus and Systems, Vol PAS-100, no. 5, May 1981, pp 2258 -2262.

[4]. V. K. Sood. "An Introduction to Forced-Commutated HVDC Inverters". Canadian Electrical Association Spring Meeting, Montréal, 24 - 26 March 1985.

[5]. Tomas Jonsson, Per-Erik Bjorklund, "Capacitor Commutated Converters for HVDC", Stockholm Power Tech, June 1995, Proceedings: Power Electronics, pp 44-51.

[6]. T.Holmgren, G.Asplund, S.Valdemarsson, P.Hidman, U.Jonsson, O.Loof, "A test installation of a self-tuned ac filter in the Konti-Scan 2 HVDC link", 1995 Stockholm Power Tech, Int. Symposium on Electric Power Engineering, Royal Institute of Technology and IEEE.

[7]. K. Sadek, M. Pereira, D.P. Brandt, A.M. Gole, A Daneshpooy "Capacitor Commutated Converter Circuit Configurations for DC Transmission", IEEE Transactions on Power Delivery, Vol. 13, October 1998, pp 1257 - 1264.

[8]. Tomas Jonsson, Per Holmberg, Thomas Tulkiewicz, "Evaluation of Classi-

cal, CCC and TCSC Converter Schemes for Long Cable Projects", presented at EPE 99, Lausanne Switzerland, September 1999.

[9]. M. Meisingset, A.M. Gole, R. Burton, O. Eide, R. Fredheim, "Impact of Capacitor Commutated Converters in AC Systems with Multiple DC Infeed", Conference Proceedings, the 13th PSCC (Power Systems Computation Conference), Vol. 1, pp. 516 - 522, Trondheim, Norway, June/July 1999.

[10]. Tsuyoshi Funaki, Kenji Matsuura, "Predictive Firing Angle Calculation for Constant Effective Margin Angle Control of CCC-HVdc", Paper PE-320-PRD(03-2000), IEEE PES (to be published in the IEEE PES Transactions).

[11]. A.M. Gole, M. Meisingset, "An Active Filter for Use at Capacitor Commutated HVDC Converters" IEEE Trans. on Power Delivery, Vol.16, No.2, April 2001.

[12]. A.Mazumder and V.K.Sood, "Comparative study of HVDC system with Capacitor Commutated Converters", ELECTRIMACS'2002 Conference, 18-21 August 2002, Montreal.

[13]. N.Ottosson and L.Kjellin, "Modular back-to back HVDC, with capacitor commutated converters (CCC)," IEE Int. Conf, AC-DC Power Transmission, 28-30 Nov. 2001, Conf. Publication No. 485, pp 55- 59.

[14]. J.Graham, D. Menzies and G.Bil, editors, "Electrical system considerations for the Argentina-Brazil 1000 MW Interconnection", CIGRE 2000 Conference, Paris, Aug 27 - Sept 2, 2000.

Chapter 7

Static Compensators: STATCOM Based On Chain-link Converters

7.1 INTRODUCTION

7.1.1 Static Var Compensator (SVC)

The SVC [1,2,3,4] has been used for reactive power compensation since the mid 1970's, firstly for arc furnace flicker compensation and then in power transmission systems. One of the first 40 MVAr SVC was installed at the Shannon Substation of the Minnesota Power and Light system in 1978. The SVC results in the following benefits [9]:

- Voltage support,
- Transient stability improvement, and
- Power system oscillation damping.

Although many versions of SVC's exist, the most common one (Figure 7-1a) usually employs (either thyristor or mechanically) switched capacitors and thyristor controlled reactors (TCRs). With appropriate coordination of the capacitor-switching and reactor-control (Figure 7-1b), the var output can be varied continuously and rapidly between capacitive/inductive values. It maintains the steady state and dynamic voltage at a bus within bounds, and has some ability to control stability [9], but not much to control active power flow. A detailed bibliography on SVC applications is found in [11].

In recent years, due to the development of GTOs, a new static compensator has been offered by manufacturers. Older papers have referred to the GTO-based system as an "Advanced Static Var Compensator (ASVC)". The functional operation of this device is, however, more similar to that of a rotating synchronous condenser, but without the slow response time and mechanical inertia, and so it was briefly known as the "Static Synchronous Condenser" (STATCON) [12,13]. However, current practice is to refer to these as STATCOMs - STATic COMpensators.

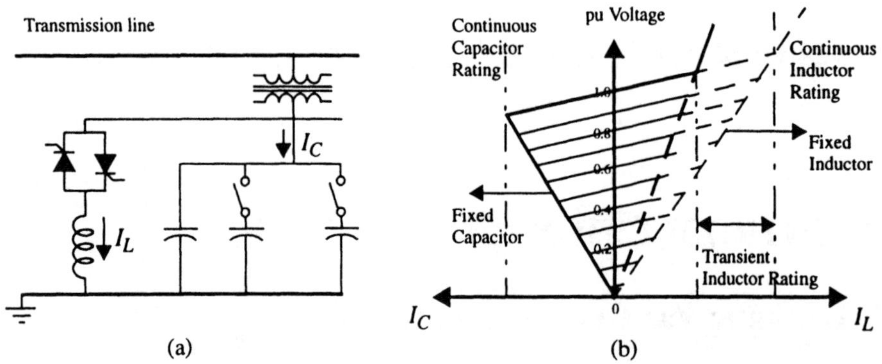

Figure 7-1: Static Var Compensator (SVC)

The ASVC is superior to the conventional SVC for the following reasons:

- Reduction in outdoor area requirement, since it replaces the voluminous capacitor/reactor banks associated with a conventional SVC,
- Improved dynamic performance and enhanced stability due to its ability to increase transiently the VAR generation,
- Improved performance at low operating voltages down to about 0.15 pu (limited only by transformer leakage), and
- Reduced need for ac filters.

The STATCOM generates a 3-phase voltage source with controllable amplitude and phase angle behind a reactance. When the ac output voltage from the inverter is higher (lower) than the bus voltage, the current flow is caused to lead (lag), and the difference in the voltage amplitudes determines how much current flows. This allows the control of reactive power.

The STATCOM (Figure 7-2a) is implemented by a 6-pulse Voltage Source Inverter (VSI) comprising GTO thyristors fed from a dc storage capacitor. Multi-pulse circuit configurations are employed to reduce the harmonic generation and to produce practically sinusoidal current. The V-I characteristic of the STATCON is shown in Figure 7-2b. The STATCOM is able to control its output current over the rated maximum capacitive or inductive range independently of ac system voltage, in contrast to the SVC that varies with the ac system voltage. Thus the STATCOM is more effective than the SVC in providing voltage support and stability improvements.

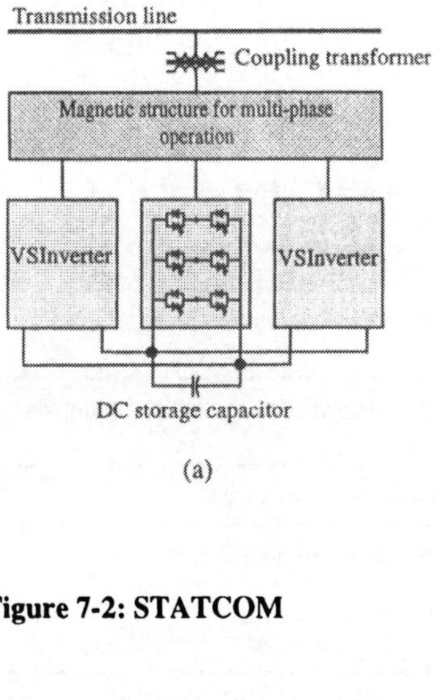

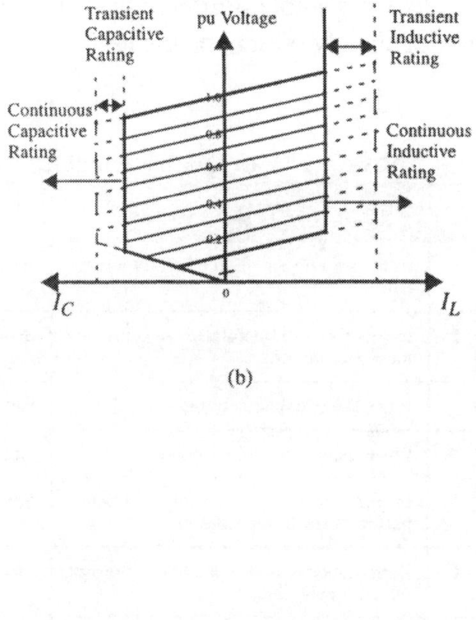

Figure 7-2: STATCOM

One difference between the STATCOM and the SVC is the performance at the limits of equipment capability. The SVC characteristic is a function of the voltage while the STATCOM can continue to produce capacitive current independent of voltage. In addition, the output current can temporarily exceed the steady-state rating. The amount and duration of the overload capability is dependent upon the thermal capacity of the GTO heat sinks and the minimum turn-off current of the GTO. With converter designs, the transient rating of the STATCOM is likely to vary from 120% to 180% of the steady state rating. Studies on the comparison of performances of SVC and STATCOM (Table 7-1) are the subject of an EPRI Project RP 3023-4.

A STATCOM version, based on IGBT switches, which is capable of operating at switching frequencies up to 2 kHz has been developed. The core parts of the plant, comprising of the IGBT valves, dc capacitors, control system and the valve cooling system are fitted into a container with a footprint of 10x20 m. The outdoor equipment is limited to heat exchangers, air-cored commutation reactors and the power transformer. A rating of ± 100 MVAr per converter is available; in case of increased rating, multiple units can be operated in parallel. The modular design makes it easily re-locatable to another site when desired to meet changing system needs. The response time of this unit is very fast (about one-quarter cycle). As a result of its high switching frequency, the plant can operate without harmonic filters, or may only require a small high-pass filter. The risk for resonant conditions is therefore negligible. Furthermore, the possibility of active filtering of harmonics already present on the network makes this an attractive choice.

Table 7-1: Comparison between a STATCOM and SVC

#	STATCOM	SVC
1	Acts as a voltage source behind a reactance	Acts as a variable susceptance
2	Insensitive to transmission system harmonic resonance	Sensitive to transmission system harmonic resonance
3	Has a larger dynamic range	Has a smaller dynamic range
4	Lower generation of harmonics	Higher generation of harmonics
5	Faster response (within ms) and better performance during transients	Somewhat slower response
6	Both inductive and capacitive regions of operation possible	Mostly capacitive region of operation
7	Can maintain a stable voltage even with a very weak ac system	Has difficulty operating with a very weak ac system
8	Can be used for small amounts of energy storage	
9	Temporary overload capability translates into improved voltage stability	

A new approach using another type of converter has been recently implemented. The advantages of this approach are that it avoids the need for a complex transformer and does not rely on the series connection of GTOs to develop higher voltage equipment. This new approach is described next.

7.2 THE CHAIN LINK CONVERTER

Each link of the Chain Link Converter (CLC) is formed of a four-arm bridge converter and a capacitor (Figure 7-3).

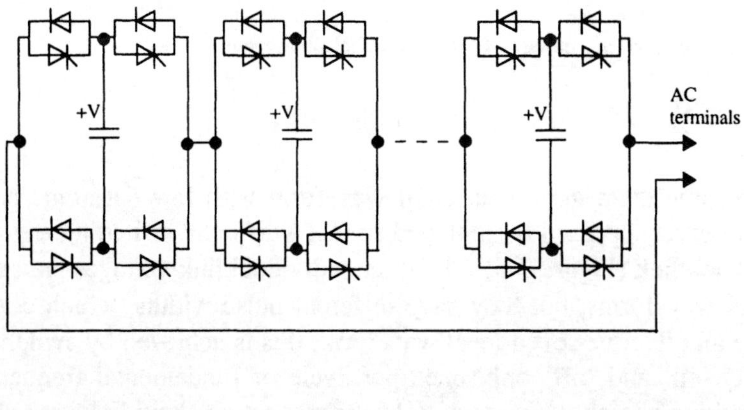

Figure 7-3: Single-phase chain circuit

Each arm is composed of a self-commutating GTO switch with a diode in anti-parallel. A number of these bridges are connected in series to form one phase of the CLC. A three phase converter would therefore be comprised of three such chains connected in either star or delta configuration.

The total chain voltage is the sum of the individual link voltages. Each pair of arms of the bridge operates as a two-way connection (Figure 7-4). At any instant, a bridge arm can connect to either the positive or negative plate of its capacitor. By suitable choice of the arm connections, each capacitor can contribute either a positive, negative or zero voltage (bypass mode) to the chain link voltage. In this fashion, the total chain, with N links connected in series, can synthesize a waveform with (2N+1) levels.

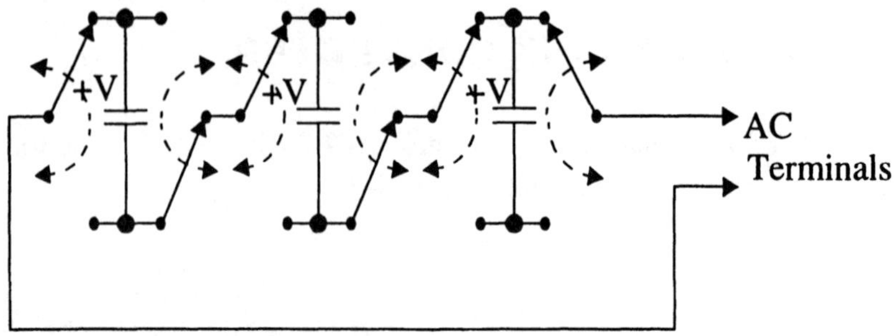

Figure 7-4: Switch analogy of three link chain converter

To synthesize a good sinusoidal waveform with low harmonic content, a multi-level approach is employed so that a different switching angle is used for each link (Figure 7-5). The three individual link voltages are still three-level waveforms, but they have different pulse widths, which combine to give an effective seven-level waveform; this is achieved by switching each GTO 'on' and 'off' only once per cycle of fundamental frequency. The switching angles α_1, α_2 and α_3 of the respective chain links are chosen so that the summed voltage is a good approximation to a sine wave.

7.2.1 Chain Link Ratings

The chain link rating is defined by the ratings of the GTOs employed. Typical commercial devices have a peak voltage rating of 4.5 kV and a peak turn-off current capability of 4 kA. In practice, de-rating of these values is necessary to withstand switching transients and provide adequate safety margins for the stresses encountered during faults and disturbances.

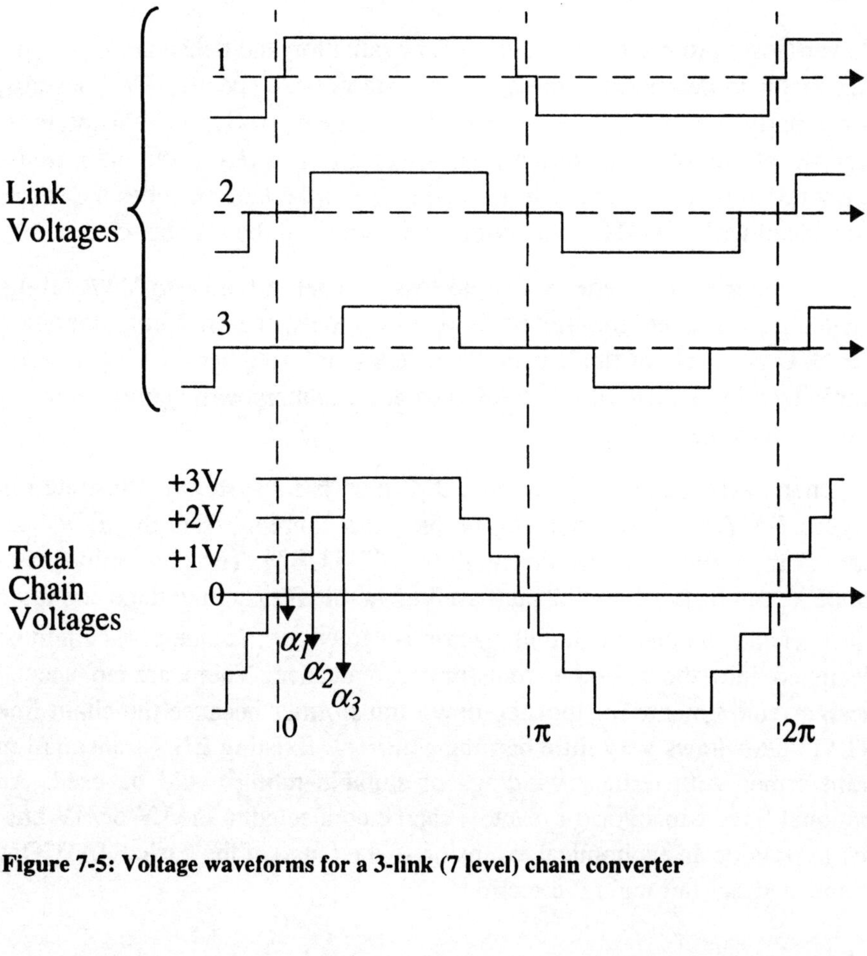

Figure 7-5: Voltage waveforms for a 3-link (7 level) chain converter

The converter VAr rating is increased by the addition of more chain links in series to raise the converter voltage; the current rating remains unaltered as it is difficult to operate GTOs in parallel. Ratings of the order of ± 100 MVAr or more in one 3-phase unit is possible using the same basic link design as a building block.

Redundancy is designed into the chain circuit converter by adding an extra link per phase; then in the event of a GTO failure, the affected chain link operates continuously in the bypass mode, until the next planned maintenance outage. Since all of the links in the chain are identical, the control system adapts itself automatically to re-optimize the switching pattern on the remaining links to minimize the generated harmonic voltage.

7.2.2 Losses

Power losses are usually capitalized for evaluation and hence are an important factor in determining the economic viability of specific SVC designs, particularly in transmission systems. For all types of STATCOM the most significant contributions to the total losses are incurred in the main transformer(s) and the power electronics equipment; the later includes the losses associated with the GTO conduction, switching and the snubber circuit.

The chain circuit converter maintains losses to below 0.65% of MVA rating. Switching losses are minimized by requiring only one switching operation per GTO per cycle of fundamental frequency, and snubber circuit losses are minimized by the use of a low loss design. Snubbers with energy recovery are also feasible.

A general arrangement of a STATCOM in an EHV system is illustrated in Figure 7-6. A stepdown transformer provides coupling from the EHV busbars to the point of connection with the STATCOM. The connection reactance X_c can be typically 0.2 pu (based on nominal system voltage and rated current) and can be an external reactor as shown, or in some cases could be designed into the effective transformer reactance. There are no special design requirements for the step-down transformer because the chain link STATCOM draws very little harmonic current. Existing EHV transmission transformer with tertiary windings of suitable rating could be used. An optional fixed capacitor (or reactor) can be connected to the LV or HV busbar to provide an economical extension to the range of the basic STATCOM in the leading (or lagging) direction.

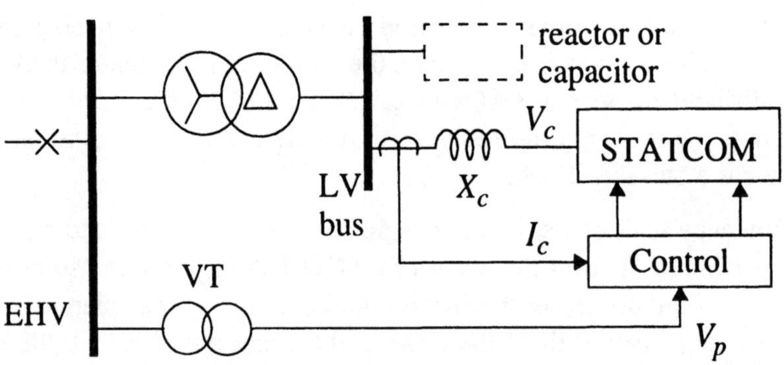

Figure 7-6: General arrangement of a STATCOM

7.3 ADVANTAGES OF CHAIN CIRCUIT STATCOM

The CLC has the following advantages in comparison to other GTO based circuit topologies and conventional thyristor SVCs [9]:

- VAr rating can be increased simply by adding chain links in series, the converter cost being in proportion to total rating,
- CLCs are essentially single phase which offers the potential for ac system phase balancing,
- There is only one GTO turn-on and turn-off switching operation per cycle, giving a low loss design,
- Low loss snubber circuit and snubber energy recovery can be implemented to minimize losses,
- Redundancy against a chain link failure can be built-in,
- Because chain links switch in sequence, the maximum instantaneous voltage excursion of the converter waveform is approximately 2 kV. Therefore, radio interference is minimized,
- Only one transformer of conventional design is required to step down from transmission voltage to the chosen STATCOM connection voltage,
- Good harmonic performance can be achieved with small or no harmonic filters. With a suitable choice of switching angles the generation of low-order harmonics (particularly triplens) can be prevented during unbalanced ac system conditions, which is impracticable with circuit based on 3-phase bridges,
- A fast response is inherent with STATCOM technology. The transient performance is enhanced by the ability to instantaneously change the output voltage by the independent control of the switching angle of each chain link and by inserting or bypassing links,
- The problem and limitations of the direct series connection of GTO's are avoided,
- Due to its constant current characteristic, a chain link STATCOM can operate down to low ac system voltage and maintain full rated leading current to support the ac system during faults,

- Space saving of around 50% relative to conventional thyristor-based SVCs of the same rating, hence the STATCOM can be a relocatable design,
- The reservoir capacitors in each link can be replaced with a battery or other suitable energy storage component giving the equipment the capability of providing real power compensation to the ac system,
- This could have application in frequency control, peak lopping, replacement of spinning reserve and starting an unenergised ac system (i.e. black start),
- A chain circuit STATCOM could be connected in series with each phase of an ac system to provide a controllable positive or negative reactance to control power flow. This could have application in the control of power flow in parallel circuits of an ac system, and
- The ability of the STATCOM converter to control self-generated harmonics could be utilized to control pre-existing harmonics on ac systems. This would give the converter an active filtering capability up to the rating of the equipment.

7.4 DESIGN FOR PRODUCTION

The first installation of a CLC based STATCOM entered service in 1999 and was the ± 75 MVAr STATCOM ordered by the National Grid Company (Figure 7-7) as part of a 0-225 MVAr relocatable SVC for initial installation at its East Claydon substation [10].

This ensures that the STATCOM operates at a low current in the float region 0-30 MVAr, to give less than 0.1% losses and consequently only a small value of loss capitalization.

The TSC was selected to provide the additional SVC capacitive output when more than about 100 MVAr is required. To avoid discontinuity in the characteristics or any tendency for the controls to hunt near to the TSC switching point, the TSC rating is less than the dynamic range of the STATCOM. When the TSC is switched off, the SVC has an inductive range, which can contribute to temporary system needs under disturbed conditions.

The total site area for the SVC equipment was 35x40 m^2. Transportable cabins accommodate the indoor-type equipment and all other equipment is designed for straightforward mounting and connection, using skids to group items together for ease of transport.

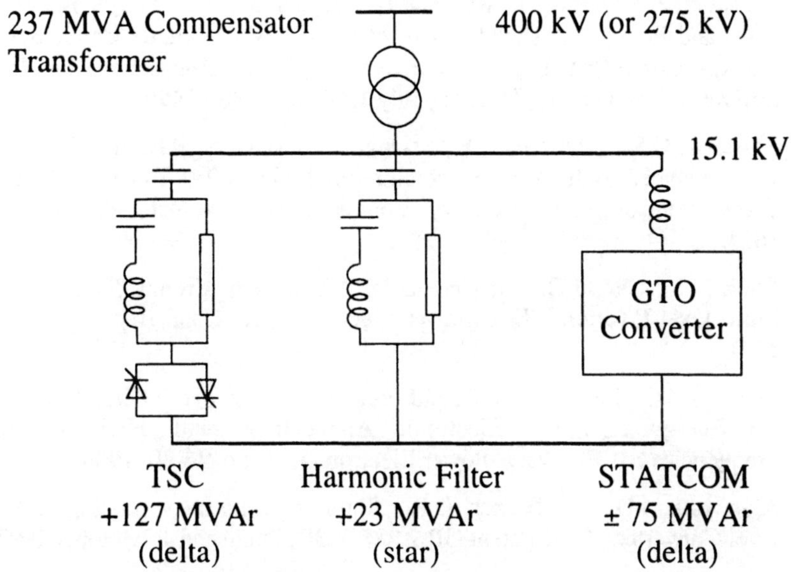

Figure 7-7: NGC East Claydon STATCOM based SVC

7.5 ACKNOWLEDGEMENTS

The support and technical assistance provided by Mr. M. Baker of Alstom in the UK is gratefully acknowledged.

7.6 REFERENCES

[1]. Young, D.J., Horwill, et al, "City Versus Country: a comparison of two types of standardised SVC for the National Grid in England," IEE 5th International Conference AC and DC Power Transmission, London, UK, pp. 248-254, 1991.

[2]. Horwill, C., Young, D.J. and Wong, K.T.G., "Design for a re-locatable tertiary connected SVC," IEE 6th International Conference AC and DC Power Transmission, UK, pp. 399-404, 1996.

[3]. Trainer, D.R., Tennakoon, S.B. and Morrison, R.E., "Analysis of GTO-

based Static VAR Compensators," IEE Proc. Electric Power Applications, 141, (6), pp. 293-302, 1994.

[4]. Schauder, C., Gernhardt, M., Stacey, E., Lemuk, T., Gyugyi, L., Cease, T.W. and Edris, A., "Development of a ±100 MVAr Static Condenser for Voltage Control of Transmission System," IEEE Transactions on Power Delivery, ,Volume: 10 , Issue: 3 , July 1995, pp.1486 - 1496.

[5]. Edwards, C.W., Mattern, K.E., Stacey, E.J., Nannery, P.R. and Gubernick, J., "Advanced Static VAR Generator Employing GTO Thyristors," IEEE Transactions on Power Delivery, Volume: 3, Issue: 4, Oct. 1988, pp.1622 - 1627.

[6]. Choi, N.S., Cho, J.G. and Cho, G.H., "A General Circuit Topology of a Multi level Inverter," IEEE Power Electron. Spec. Conf. Rec., pp. 96-103, 1991.

[7]. Bendien, J.C., Broeck, H.V.D. and Fregien, G., "Recovery Circuit for Snubber Energy in Power Electronic Applications, with High Switching Frequencies," IEEE Trans. Power Electron, 3, (1) pp. 26-30, 1988.

[8]. Ainsworth, J.D. and Trainer, D.R., "Improvements in or relating to Multi level Converters," UK Patent GB 2 285 523B, Published 15 October 1997.

[9]. FACTS Overview IEEE Catalogue 95 TP108, Produced by CIGRE and IEEE, April, 1995.

[10]. D.Hanson, "A transmission SVC for National Grid Company plc, Incorporating a ± 75 MVAr STATCOM", IEE Colloqium on Flexible AC Transmission Systems, London, Nov. 1998.

[11]. I.A. Erinmez, Editor, "Static Var Compensators", Working Group 38-01, Task Force No.2 on SVC, CIGRE, 1986.

[12]. I.A. Erinmez, and A.Foss, Editor, "Static Synchronous Compensator (STATCOM)", Working Group 14-19, CIGRE Study Committee 14, Document 144, August 1999.

[13]. Arindam Ghosh, Gerard Ledwich, "Power Quality Enhancement Using Custom Power Devices", August 2002, ISBN 1-4020-7180-9, Book published by Kluwer Academic Press.

Chapter 8

HVDC Systems Using Voltage Source Converters

8.1 INTRODUCTION

Traditional HVDC transmission used current source converters with line-commutated thyristor switches. The dual voltage source converters could not be used until the early 1990s due to the lack of commercially available self-commutating switches (i.e. GTOs and IGBTs). The availability, first, of commercially available high power GTO switches and then, secondly, of IGBT switches changed the scene. The use of VSC has taken off in the past 10 years, and new applications are being announced frequently. However, due to the higher cost of a VSC compared to a CSC on a per kW basis, the VSC applications are reserved for niche applications which could not be served by CSCs. As the cost of VSCs comes down, the number and variety of applications will increase due to the increased benefits offered by such converters. A comparison between CSCs and VSCs is made in Table 8-1.

Two prominent manufacturers refer to the new technology of dc transmission using VSCs by commercial trade names such as **HVDC Light** [1] and **HVDC^{PLUS}** (the "plus" stands for Power Link Universal Systems) [3]. The recent interest in this new technology has grown due to a number of factors.

Deregulation in the electric power industry, coupled with continued load growth and the difficulty of obtaining RoWs for new transmission lines, implies that existing transmission system assets be utilized efficiently and closer to their thermal limits. As the existing ac lines are loaded nearer to their thermal limits, losses will increase, power quality will deteriorate and network stability will be negatively impacted. This will necessitate modifications of existing transmission assets to increase power density on existing

RoWs. An approach using dc transmission based on VSCs has the potential to aid in the solution to these anticipated problems. The advantages of this alternative approach are:

- The VSCs provide an independent control of both active and reactive power. This feature is very attractive in a city centre (with a significant number of underground ac cables) where reactive power control is both complicated and expensive to implement,
- Compact, modular, standardized construction of the convertor permits a factory-tested station which can be rapidly installed/commissioned on site. Furthermore, the station size can be expanded in a staggered manner to suit the system growth, and
- The new XLPE (Cross Linked Poly-Ethylene) extruded polymer dc cables can fit into existing ac cable ducts or RoWs and can provide almost a 50% power transfer capacity increase at the same time.

Table 8-1: Comparison of current-source versus voltage-source converters

Current Source Converters	Voltage Source Converters
Uses inductor L for dc side energy storage Uses capacitor C for ac side energy storage	Uses capacitor C for dc side energy storage Uses inductor L for ac side energy storage
Maintains constant dc current	Maintains constant dc voltage
Fast accurate control	Slower control
Higher losses	More efficient
Larger than 300 MW	Less than 300 MW
More fault tolerant and more reliable	Less fault tolerant and less reliable
Simpler controls	Complexity of control system is increased
Not easily expandable in series	Easily expanded in parallel for increased rating

8.2 BASIC ELEMENTS OF HVDC USING VSCs

The approach is being pushed by advances in two basic technologies:

- Voltage Source Converters, and
- XLPE cables for HVDC transmission.

8.2.1 Voltage Source Converters

With present day IGBT high-power switches, the VSC rating can be extended to about ±150 kV, 300 MW in a bipolar link. Furthermore, the VSC permits connection to weak or even dead networks.

The control method is based on pulse width modulation (PWM) techniques which enables the flexible and independent control of both active and reactive power and the ability to limit the generation of low-order harmonics. This provides for a high level of power quality.

The VSC is of a compact, modular design which is pre-assembled in a container and is completely tested in a factory. The container can be re-located to a site and connected to the system for rapid, cost-effective installation and commissioning. Since the design is modular, the installation can be planned in stages to accommodate future load growth.

8.2.2 The XLPE Cables

In ac transmission, cable technology has changed from paper insulated cables to XLPE (Cross Linked Poly-Ethylene) polymer extruded cables. However, the use of extruded cables for dc was delayed for many years due to two reasons:

- The existence of space charges in the insulation leading to uncontrolled local high electric fields causing dielectric breakdowns, and
- Uneven stress distribution due to temperature dependent resistivity causing over stress in the outer part of the insulation.

These problems have now been resolved resulting in new dc cables which have an insulation of extruded polymer. The insulation system is triple extruded, i.e. the conductor screen, the insulation and the insulation screen are all extruded simultaneously. The insulation provides a robust construction. Hence the cable is easy to handle and install in the following applications:

- Directly ploughed underground,
- Insulated aerial, and
- Severe submarine locations (deep water and a rough sea-bed).

These new XLPE cables now replace the older paper insulated cables, Low Pressure Oil Filled (LPOF) cables or Mass Impregnated Non Draining (MIND) cables which were previously used for HVDC transmission. The LPOF cables needed auxiliary equipment to maintain the oil pressure, could not be easily installed and had environmental oil spill concerns associated with them. The MIND cables had limitations in their operating conductor temperature. Furthermore, paper insulated cables were not feasible for aerial applications because of their sensitivity to repeated bending.

8.2.2.1 Comparing AC-DC Cables

When compared to ac cables, dc cables are preferred for power transmission as they provide a longer lifetime than ac cables due to the charging current requirement. Furthermore, ac cables cannot be used for a distance of over 50 kms due to charging current requirements. The new dc cables have successfully passed the 20 kV/mm stress lifetime and type tests. Magnetic fields are eliminated since the dc cables are laid in pairs for bipolar operation with anti-parallel dc currents.

8.3 VOLTAGE SOURCE CONVERTER

A VSC is composed of a 6-pulse bridge equipped with self-commutating switches (either GTOs or IGBTs) and diodes connected in anti-parallel (Figure 8-1). To achieve the required rating, a number of switches are connected in series to build one valve. GTO valves allow higher currents but lower dc voltage than comparable IGBT valves.

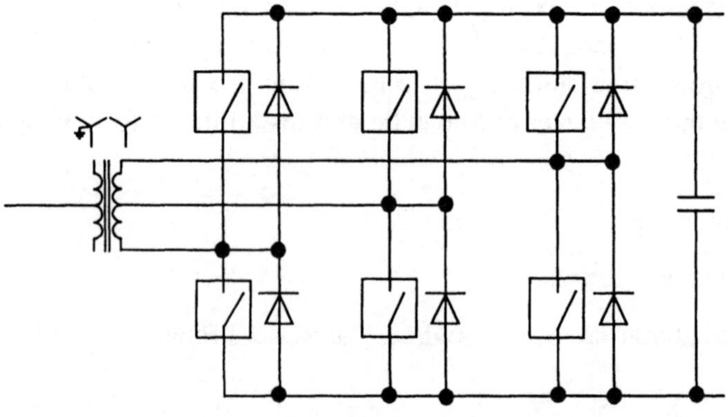

Figure 8-1: Voltage Source Converter bridge

For longer distance transmission, since it is desirable to maintain a high transmission voltage to reduce losses, IGBT valves are preferred to GTO valves for such applications. With the present-day technology, IGBT valves can block up to 150 kV. A VSC equipped with such valves can carry up to 1000 A (rms) ac line current. This results in a designed power rating of approximately 150 MVA per VSC. Thus, a bipolar system can easily be built up to handle power levels of about 300 MW. At the lower end of the scale, an economical design can be built for just a few MWs.

In fact, one of the first installations of this technology was the test transmission at Hellsjon in March 1997 where a 3 MW, ±10 kV, dc transmission over a distance of 10 km using a dc cable was implemented. Some of the other recent installations of this technology are shown in Table 8-2.

8.3.1 Operating Principles Of A VSC

The basic operating principle of a VSC is shown in Figure 8-2. Its function is to convert the dc voltage of the storage capacitor into an ac current. The polarity of the dc voltage of the converter is defined by the polarity of the diode bridge. The VSC valves can be switched on at any time by appropriate gate voltages. However, if one valve is switched on, then its complementary valve must have been switched off previously to prevent a short circuit of the storage capacitor. Alternate switching of the valves connected to one phase module successively connect the ac terminal of the VSC to either the positive or negative plates of the dc capacitor. This results in a square wave ac voltage comprising two voltage levels: $+V_{dc}/2$ and $-V_{dc}/2$. Such a VSC is therefore referred to as a 2-level converter [3].

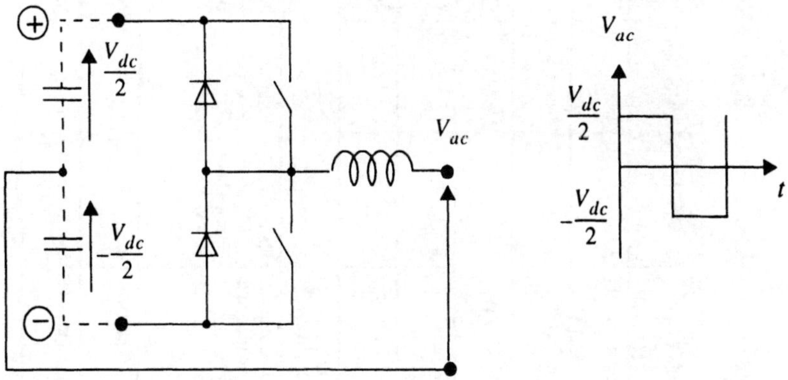

Figure 8-2: Principle of VSC

Table 8-2: Installed HVDC projects based on VSCs

Scheme	Hallsjon	Hagfors	Gotland	Directlink	Tjaerebog	Eagle Pass	Mosels Tahlwerke	Cross Sound Cable	Murray Link
Commissioning Year	1997	1999	1999	2000	2000	2000	2000	2001	2002
Power Transmitted, MW	3	NA	50	180	7.2	36 MW	NA	330	200 MW
Direct Voltage, kV	± 10	NA	± 80	± 80	± 9	± 15.9 kV	NA	± 150	± 150 kV
Converters per station	1	1	1	3	1	1 dual or 2 single	1 single	1	1
Direct Current, A	150	NA	360	375	358	1130	NA	1200	-
Reactive power supply, MVar	±3	± 22	± 30	± 75	-3 / +4	± 36	0-38 controllable	± 75	+140 / -150 MVAr
Converter station location and AC grid	Hallsjon, 10 kV, 50 Hz Grangesberg, 10 kV, 50 Hz	Hagfors, 36 kV, 50 Hz	Nas, 77 kV, 50 Hz Backs, 77 kV, 50 Hz	Terranora, 110 kV Mullumbimby 132 kV, 50 Hz	Enge, 10.5 kV, 50 Hz Tjaereborg 10,5 kV, 50 Hz	Eagle Pass, 138 kV, 60 Hz Intertie USA & Mexico 60 Hz	Trier, 20 kV, 50 Hz	New Haven, 345 kV, 60 Hz Shoreham, 138 kV, 60 Hz	Berri, 132 kV Red, 220 kV
Cable Length km	0.2	NA	70	65	4.4	0 (B-B)	NA	40 km	180 km
AC grids at both ends	Synchronous		Synchronous	Asynchronous	Normally synchronous	138 kV Asynchronous	NA	Synchronous	Asynchronous

Recent practices have extended the principle to multi-level converters by the use of capacitors and diodes to increase the number of levels and clamp the voltages for the different levels. This provides for improved waveform quality. As an example, a 3-level converter is shown in Figure 8-3 and its principle of operation is shown in Figure 8-4.

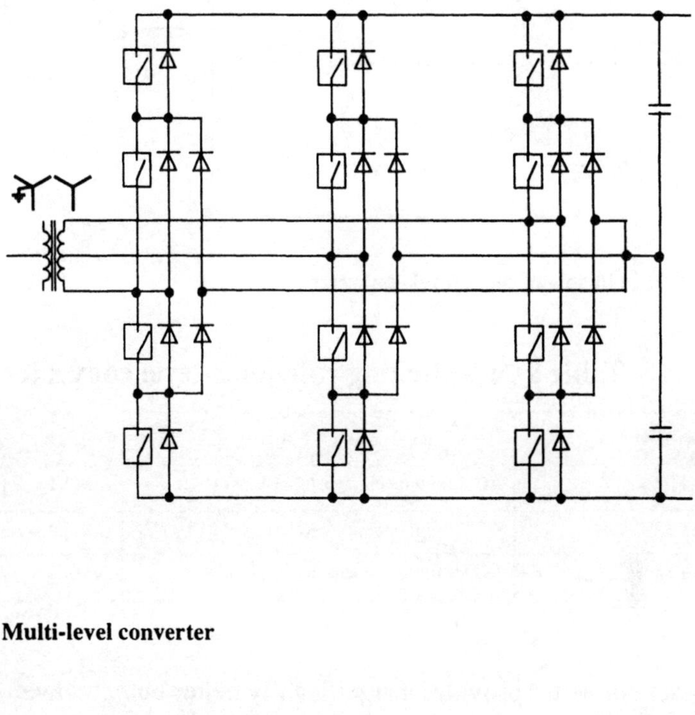

Figure 8-3: Multi-level converter

The 3-level converter comprises of 4 valves in one arm of the converter. The switching rule is that only two valves that are directly connected can be switched ON at any one time i.e. S1 & S2, or S2 & S3, or S3 & S4. Switching on S1 & S2 connects the ac terminal AC to the positive terminal P, S2 & S3 connects to the Mid-Point (MP) terminal via the clamping diodes, S3 & S4 connects to the negative terminal N of the dc system (Table 8-3). Thus, the resulting voltage on the ac terminal comprises three voltage levels instead of two, as in the case of the 2-level converter.

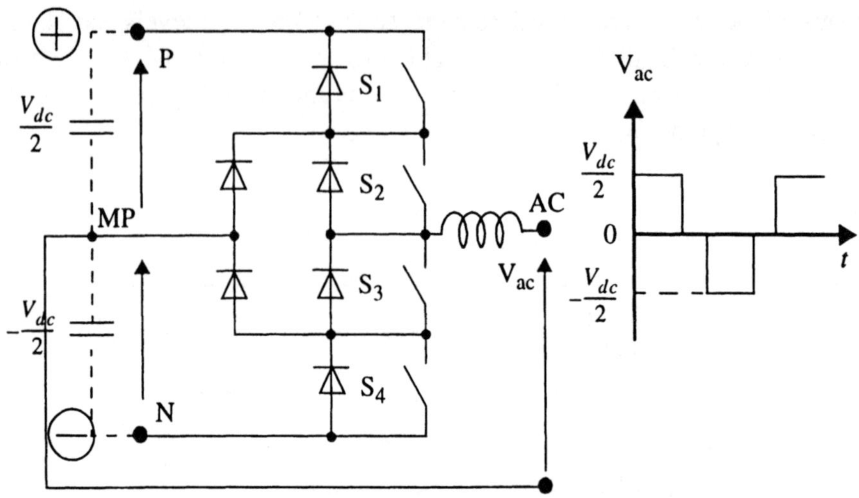

Figure 8-4: Principle of the 3-level converter

Table 8-3: Switching rule for 3-level converter

Switches ON	AC Output	Level
S1 and S2	AC terminal connects to terminal P	Level 1 = $V_{dc}/2$
S2 and S3	AC terminal connects to terminal MP	Level 2 = zero
S3 and S4	AC terminal connects to terminal N	Level 3 = - $V_{dc}/2$

A 3-level converter provides a significantly better output waveform quality as far as the Total Harmonic Distortion (THD) voltage is concerned. However, the more complex converter design results in a larger footprint and higher costs making the 2-level converter technology the preferred solution for the moment.

8.3.1.1 Design Of Control Systems

By using Pulse Width Modulation (PWM) techniques with a high switching frequency in the 1-2 kHz range, the wave shape of the converter ac voltage output can be controlled to be almost sinusoidal with the aid of a small high frequency filter. Furthermore, since the generated output voltage can be virtually at any angle and amplitude with respect to the bus voltage, it is possible to control independently both the active and reactive power flow.

The VSC acts a motor/generator with almost no inertia and therefore, it can control active/reactive power almost instantaneously. Also, since it has virtually no inertia, it does not contribute to the short circuit current.

8.3.2 Design Considerations

8.3.2.1 Steady State Characteristics [3]

Figure 8-5a shows a simplified single line diagram to calculate the fundamental load flow between the ac system and the converter. The ac system as well as the converter are considered as voltage sources. These voltage sources are connected via the transformer, represented as an ideal transformer and a leakage impedance. While the magnitude and phase angle of the ac system voltage V_N is determined by the load flow of the ac system, the magnitude and phase angle of the converter voltage V_{con} can be varied by the converter control. According to Kirchhoff's law, controlling V_{con} relatively to V_N results in a corresponding voltage ΔV across the transformer leakage impedance X_T. Neglecting losses, the current I through X_T lags ΔV by 90°el. As a consequence, controlling V_{con} determines the current I.

Figure 8-5b shows the relations between voltages and currents for an operating point as an example. The current I can be split up regarding V'_N into one component that is in parallel and into another component that is in quadrature.

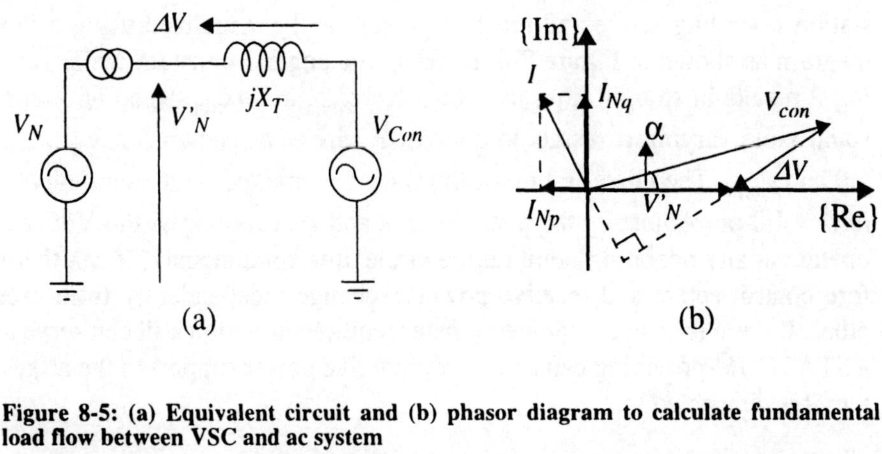

Figure 8-5: (a) Equivalent circuit and (b) phasor diagram to calculate fundamental load flow between VSC and ac system

The active and reactive power exchange as seen from the ac system terminals of the transformer can be calculated according to the following formulas:

$$p_N = \frac{v_N^2}{\theta_x}\lambda\sin(\alpha) \quad \text{and} \quad q_N = \frac{v_N^2}{\theta_x}(1-(\lambda+\cos(\alpha))) \qquad (8\text{-}1)$$

with

$$v_N = \frac{V_N}{V_{Nn}}\;; p_N = \frac{P_N}{S_{bT}}\;; q_N = \frac{Q_N}{S_{bT}}\;; \theta_x = X_T\frac{S_{bt}}{V^2_n}\;; \lambda = \frac{V_{con}}{V_N}$$

V_N:	present magnitude of the ac system voltage
V_{Nn}:	magnitude of the ac system nominal voltage
V_{con}:	present magnitude of the converter ac voltage
p_N:	present active power exchange as seen from the ac system
q_N:	present reactive power exchange as seen from the ac system
S_{bT}:	rated apparent power of the transformer

Considering a certain ac system voltage V_N, the operating range of a VSC station regarding active and reactive power can be visualized using a PQ-diagram as shown in Figure 8-6. Keeping the angle α constant while varying λ results in straight lines as shown for α_{max} and α_{min}. Keeping λ constant while varying α results in concentric circles as shown for $\lambda = \lambda_{max}$, 1.0 and λ_{min}. The power rating of the station is marked by the circle $|pN + jqN| = 1.0$ pu. Adjusting the parameters λ and α accordingly, the VSC can operate at any operating point on the circle area continuously. It can therefore control active and reactive power exchange independently from each other. If there is no active power transfer required, a station still can serve as a STATCOM providing capacitive or inductive power support to the ac system it is connected to.

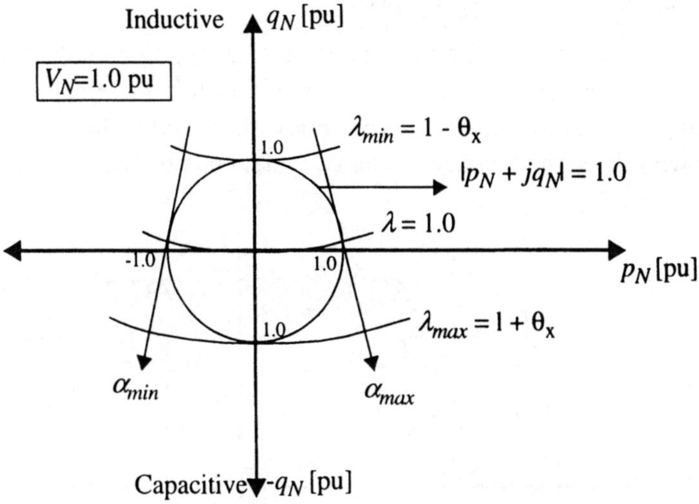

Figure 8-6: PQ diagram for the VSC [3]

8.4 APPLICATIONS

Some areas where the VSC technology will be useful are:

- In environmentally sensitive locations, i.e. city centres,
- Infeed of small scale generation, i.e. wind or other new generation systems,
- Power supply to remote loads (i.e. islands with dead loads), and
- Asynchronous inter-connections i.e. Directlink.

Examples of some pioneering applications using this technology are provided next.

8.4.1 In Environmentally Sensitive Locations, i.e. City Centres

Adding new transmission capacity by ac lines into city centres is costly and in many cases the permits for new ROWs are difficult to get. A dc cable needs less space than an ac overhead line and can carry more power than an ac cable and is therefore, many times the only practical solution, should the city centre need more power [1].

The output from a VSC converter always has the same polarity. This makes it easy to use a building block in a multiterminal system. To a dc bus with fixed polarity, any number of VSC converters, and by that a meshed dc system with the same topology as an ac system can be built (Figure 8-7).

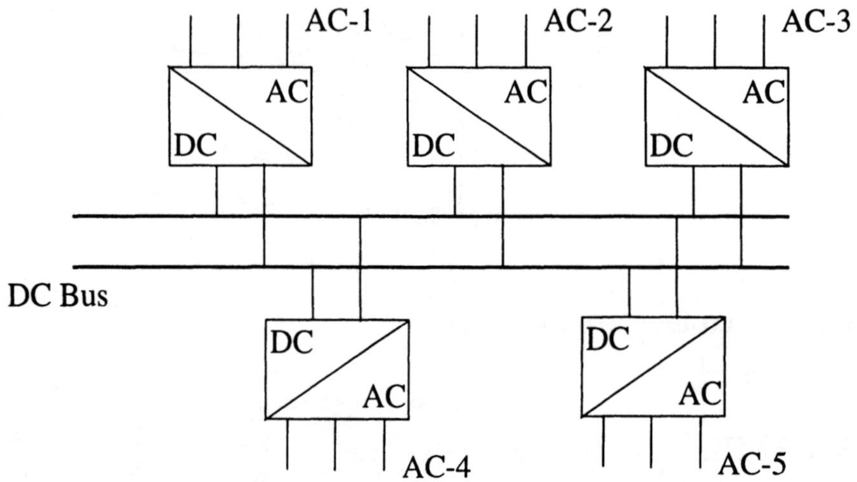

Figure 8-7: Multi-Terminal HVDC system for city centres [1]

The implementation of a dc distribution system for city centres is still awaiting further developments in equipments such as electronic transformers and current limiting devices to maximize flexibility and optimize the utilization of fixed assets.

8.4.2 Infeeds Of Small Scale Renewable

DC technology is advantageous for feeding power from other small scale generations to a main ac grid. It can be used for photo voltaic power transmission, small hydro schemes or for power from off-shore gas turbine generators to the ac grid via submarine dc cables (Figure 8-8). One advantage with dc transmission is that generators are free to operate at a different frequency than the network or even at variable frequency.

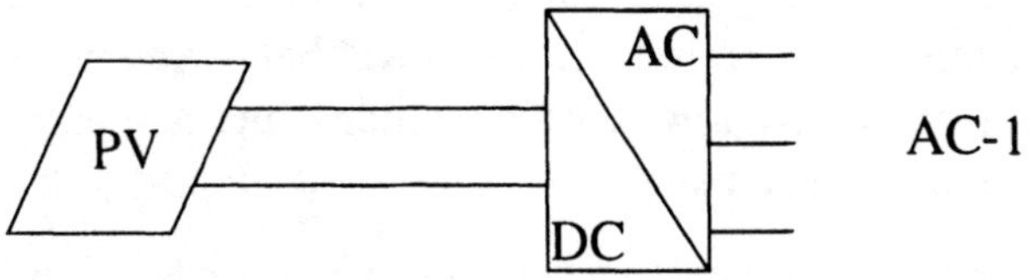

(a) Photo voltaic system connected to ac grid

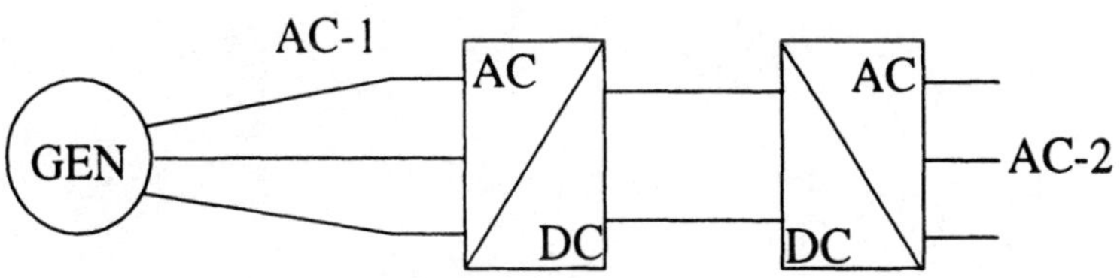

(b) Wind or small scale hydro generation connected to ac grid

Figure 8-8: Small scale generations to a main ac grid

8.4.3 Power From Wind Farms

Small scale, environmentally friendly, wind power generation systems are often located in areas far away from the main ac grid. Such schemes are often built as a group of a few units in a farm. A growing wind farm, or an area with several wind farms, often represents a real challenge to network planning and operation. A growing generation complex needs a flexible, and yet cost effective infeed to the main grid. It also needs an efficient means of controlling the ac voltage, the reactive power production, and stability of the grid with a growing percentage of highly variable wind power generation.

Although wind power is very complex form of generation, dc technology based on VSCs is highly suitable as a transmission link from wind farms to the main ac grid. The technology is flexible, i.e. it can be easily expanded by adding new units when the wind farms are expanded. Its ability to control the production of reactive power for the generator, independently of the transmission of active power, adds considerable stability margins to the ac grid even during highly variable generating and loading conditions. In addition, dc technology has the ability to eliminate voltage flicker due to wind turbulence, including the flicker caused by tower shadow.

Through control of the wind turbine bus frequency, the wind turbine rotor speed can be changed to match the needs of the ac grid. For instance, the power infeed can be reduced to a level below the maximum available production or, during low wind conditions, the active power from the wind power units can be increased for transient support of the ac grid.

There are several ways to connect one or more wind farms to an ac grid with dc technology. On the generating side, one converter station can be directly connected to the transformers of the individual units, see Figure 8-9.

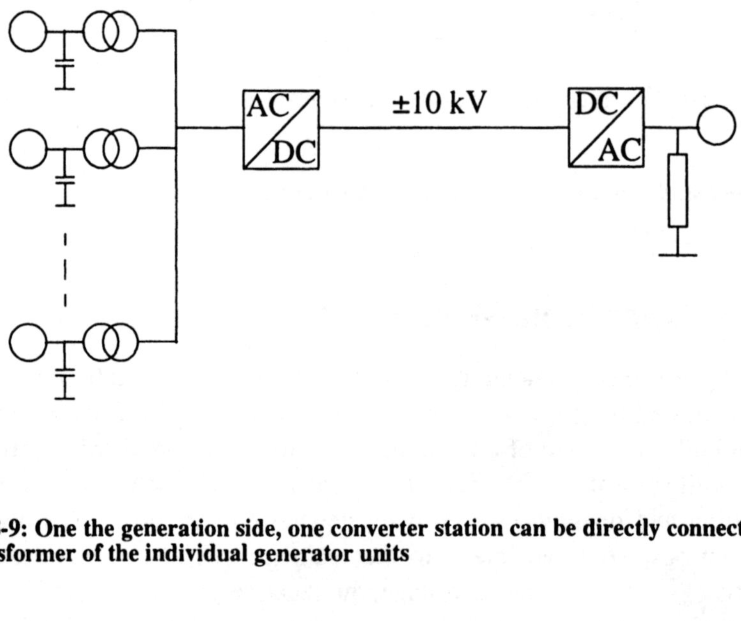

Figure 8-9: One the generation side, one converter station can be directly connected to the transformer of the individual generator units

Large wind farms can also be divided into groups directly connected to separate converters. A medium voltage dc grid then feeds the power to the main dc transmission station, which is operated in dc/dc mode, as shown in Figure 8-10.

The control strategies can include several or all of the following options: Control of the wind turbine bus voltage and frequency as well as voltage or reactive power control at the common connection point to the ac grid.

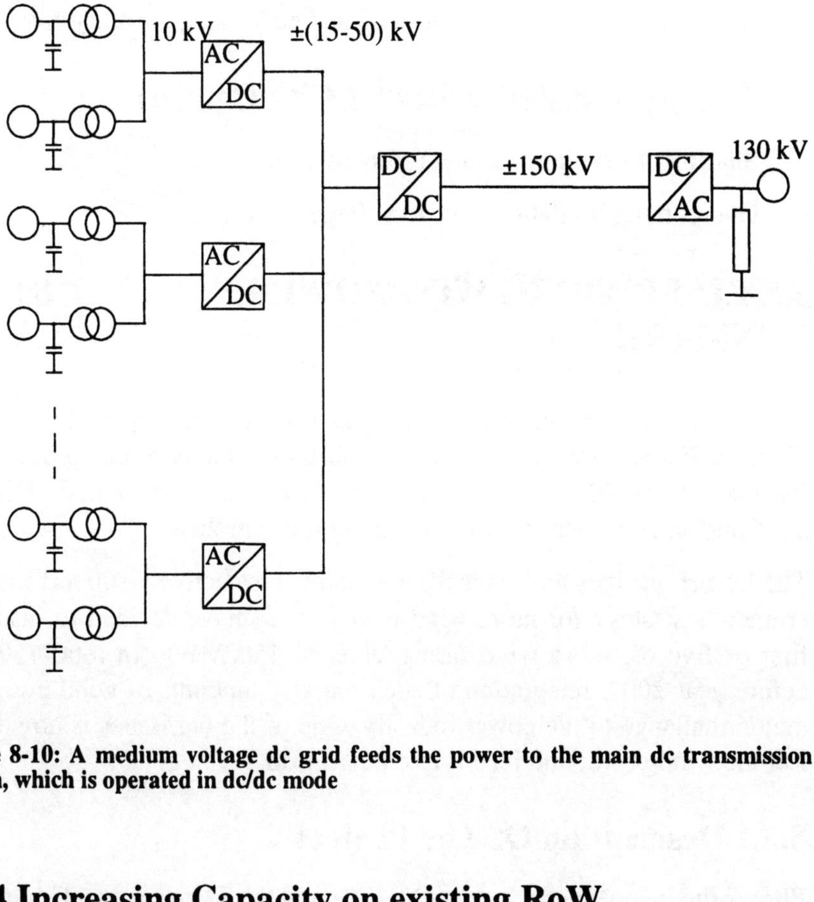

Figure 8-10: A medium voltage dc grid feeds the power to the main dc transmission station, which is operated in dc/dc mode

8.4.4 Increasing Capacity on existing RoW

- **Converting ac to dc**

  Previously, to increase power transmission capacity ac lines were reinforced by re-conductoring and voltage upgrades. In the future, the existing ac lines could be converted to dc transmission which would increase capacity by some 50%. The costs of conversion would include the two converter stations and the replacement of ac insulators.

- **Adding capacity with dc cables**

  Since dc cables can carry more power than ac cables, it would be more efficient to use them for the limited cable ducts into city centres and thereby increasing the power density. The robustness of the new dc cables would also facilitate the installation process.

- **Control of power flow**

  This would be possible by having an ac-dc system in parallel, where the dc system would assist in damping and stabilizing the ac system.

8.4.5 Improved Reliability Of City Centres

- Sharing of generation system reserves, and
- Overcoming limitations due to voltage stability.

8.5 TJAEREBORG WINDPOWER PROJECT IN DENMARK

A commission with members from the Danish Ministry of Energy, The National Forest, Nature Agency and the power industry has proposed the installation of 4,000 MW off-shore wind power, 2,250 MW before the year 2015 and an additional 1,750 MW before the year 2030.

The Danish utilities and transmission companies actively support the Government's strategy for more wind power. They have decided to install the first of five off-shore wind farms, each of 150 MW - in total 750 MW, before year 2007. Integration of such massive amounts of wind power is a major challenge to the power industry. One of the big issues is how to collect the wind power and feed it into the ac grid.

8.5.1 Description Of The Project

Eltra is the Independent System Operator and the Transmission Company in western Denmark. In 1998 Eltra decided to investigate the use of the new dc technology based on VSCs for connection of wind power to Eltra's transmission grid.

The purpose of the Tjaereborg (close to Esbjerg) project is to test and demonstrate on a small scale the application of this new technology for connection of large-scale off-shore wind farms greater than 100 MW and with distances above 50 km from the coastline. The project was released and started in March 1999 and was taken over for tests and demonstration in December 2000.

The Tjaereborg wind farm consists of 4 wind turbines of different types and makes (Figure 8-11), with a total generation of 6.5 MW. The dc-cable is laid in parallel with the existing ac cable, which make it possible to operate the wind farm in three different operational modes:

- Via the ac cable only,
- Via the dc cable only, or
- Via the dc and ac cables in parallel.

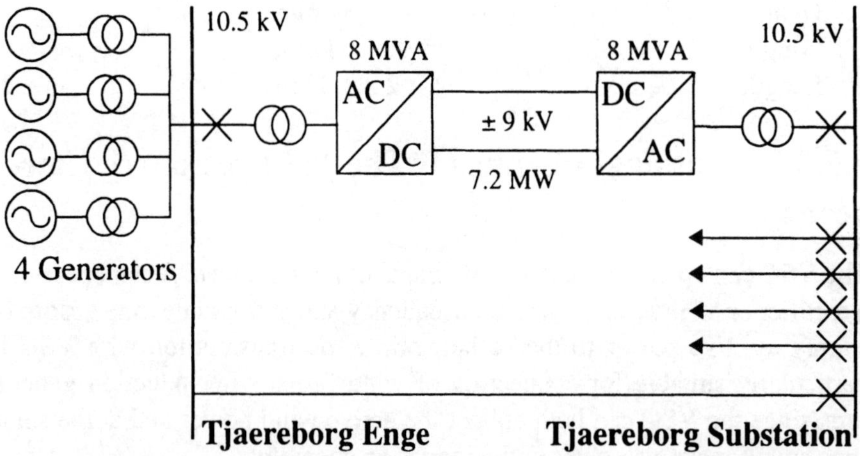

Figure 8-11: Single line diagram of Tjaereborg Substation

Operation of the wind farm in these 3 different modes makes the demonstration project suitable for a comparative investigation of the various operation modes.

With a dc feeder there are no critical cable lengths and therefore, no technical limits for how far from the coastline an off-shore wind farm can be located. Also, the frequency in the wind farm is decoupled from the network frequency and voltage, making it possible to vary frequency and voltage in the wind farm continuously and independently of the transmission grid.

8.5.2 Main Data

Rated power	8 MVA/7.2 MW
Switching frequency	1950 Hz
DC voltage	± 9 kV
DC current	358 A

AC - network	10.5 kV

Converter

Converter type	2 level VSC
Valve	Series connected IGBT

DC Cable

Aluminium conductor	240 mm^2
Diameter	34 mm
Weight	1.6 kg/m
Length	2 x 4.3 km

8.5.3 Operational Regime Of The Voltage Source Converter

The VSC can operate in all four quadrants of the PQ plane. It can operate as a rectifier or an inverter at variable frequency and at the same time absorb or supply reactive power to the ac network. A dc transmission with VSC is particularly suitable for connection of wind farms with induction generators, since the VSC can both collect the active wind power and at the same time supply reactive power to the induction generators.

8.5.4 Power Quality

- The ac side is provided with two filters tuned for the switching frequency and its harmonics,
- The Total Harmonic Distortion (THD) is less than 5% and each of the individual harmonics is less than 1%, and
- Telephone Interference Factor (TIF) is better that 50.

8.5.5 Control System

The control system used is a microprocessor based system. The design enables a very compact control system, and is built in a modular way to allow easy adaptation of additional control functions in the future. The control system has an open architecture and uses state of the art technology with industrial standard local area networks and serial communication busies. The control functions are implemented in one computer in the control cubicle and this cubicle is then duplicated to create a fully redundant system with high reliability.

Special features:

- Variable frequency operations within the frequency range 30-65 Hz in isolated operation mode,
- The converter can continuously regulate both the frequency and voltage in the isolated network operation, i.e. only the dc feeder is switched in,
- Automatic switchover between the ac feeder and the dc feeders for flexibility.

8.5.6 DC Cable

The cable insulation is made of an extruded polymer that is particularly resistant to dc voltage. Polymeric cables are a preferred choice, mainly because of their excellent mechanical strength, flexibility and low weight. The cables are ploughed into the ground by a tractor equipped with a cable plough. This makes installation fast, economical and will also meet environmental demands.

8.5.7 Building

All equipment is housed in a building with wood/steel panel walls and steel roofing which reduces the audible noise and the electromagnetic emission levels. The building is designed to blend into its surroundings.

8.5.8 Performed Tests On Site

- Start/stop of wind turbines at low and high wind velocity,
- Isolate operation by disconnecting the ac network at the sending end,
- Parallel operation of ac-dc link, and
- Varying frequency output from the converter in the wind farm.

8.5.9 Advantages

The following characteristics of the dc transmission based on VSCs make it a preferred solution for infeed of wind power in power systems:

- Variable frequency operation for optimizing power from the wind mills,
- Control of reactive power/ac voltage at point of in feed,
- Providing reactive power to the generators, and

- Use of variable frequency for spinning reserve.

8.6 POWER SUPPLY TO REMOTE LOCATIONS (i.e. ISLANDS)

Since the grid connection is often just too expensive for a remote load such as an island, power supply is often provided by a diesel generating plant. Such plants suffer from the following drawbacks:

- Environmentally unfriendly,
- Maintenance costs are high,
- Expensive transportation of diesel fuel to the remote location, and
- Noise pollution can be a problem if the plant is located close to communities.

A low-power dc transmission based on VSCs can often provide an efficient, environmentally friendly and stable power source to the remote location via dc cables. The environmental advantages are obvious, but there are other economical and technical benefits too. The maintenance cost of a dc transmission system are very low. Furthermore, since the dc system does not contribute to the fault current in the remote network, it puts lower demands on most of the breakers and may even eliminate the need for some of them.

8.6.1 The Gotland Island System

Gotland is an island situated in the Baltic Sea, some 90 km east of the Swedish mainland. The push for renewable forms of energy has brought wind power mills into focus on southern Gotland with need for additional transmission capacity, as well as for a better means to maintain a good power quality. It is a known fact that wind power mills are subjected to variable operating conditions which result in flicker and in variations of active and reactive power.

In 1997, the utility on Gotland installed a 50 MW dc link based on VSCs (Figure 8-12) located at the terminal stations of Visby and Nasudden (some 70 km south of Visby). The converter equipment was contained in compact container modules (Figure 8-13) which were factory tested. This led to a low-cost, environmentally friendly solution to the bipolar dc link which does not require electrodes. Two extruded cables (diameter 43 mm, 2 kg/m, aluminium conductor 340 mm^2) at ± 80 kV, ploughed into the ground close to each other, are used to connect the two terminal stations.

The dc link gives provides control characteristics for the converters to manage the power transfer, improve the voltage and maintain the frequency in the connected stations and thereby controlling the windmill generated flicker.

This flexibility and control options were the main reasons for the choice of dc transmission based on VSCs for Gotland. The power from a rapidly growing complex of wind farms on the southern part of the island is transmitted to the main load areas in the northern parts of the island by dc cables. The cables run in parallel to an existing ac line.

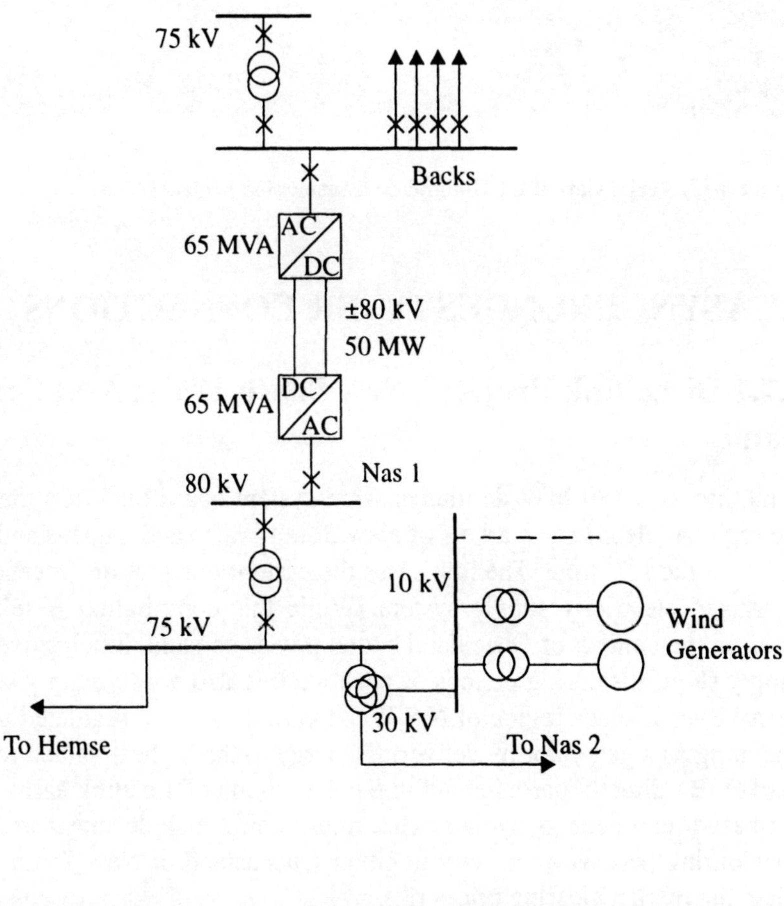

Figure 8-12: Single line diagram of Gotland dc transmission project

Figure 8-13: Aerial view of the Gotland dc transmission project [ABB]

8.7 ASYNCHRONOUS INTER-CONNECTIONS

8.7.1 Directlink Project - New South Wales And Queensland

Directlink is a 180 MW dc transmission system based on VSCs that links the regional electricity markets of New South Wales and Queensland, Australia for the first time. The link eases the current pressure on Queensland's overtaxed electricity supply system. While this contribution is relatively small in the context of Queenland's total power demand, it helps overcome supply shortfalls during periods of peak load. It also boosts supply services to the Tweed Heads region of N.S.W. Directlink is a non-regulated project, operating as a generator by delivering energy to the highest valued regional market. By directly participating in the spot market Directlink earns a market-based return for its owners. That return could include substantial revenues during periods of scarcity in either Queensland or New South Wales, when the market clearing prices rise.

8.7.2 Main System Components

The Directlink transmission system is composed of:

- Converter stations located in Terranora and Mullumbimby. Each station is rated at 195 MVA and consists of three independent 65 MVA converters (Figures 8-14 and 8-15). Other technical data is provided in the figures, and
- A total of 354 km of XLPE cables have been installed to connect the two converter stations. Two cables are used for each pair of converters using already existing rights of way for its entire 59 km route.

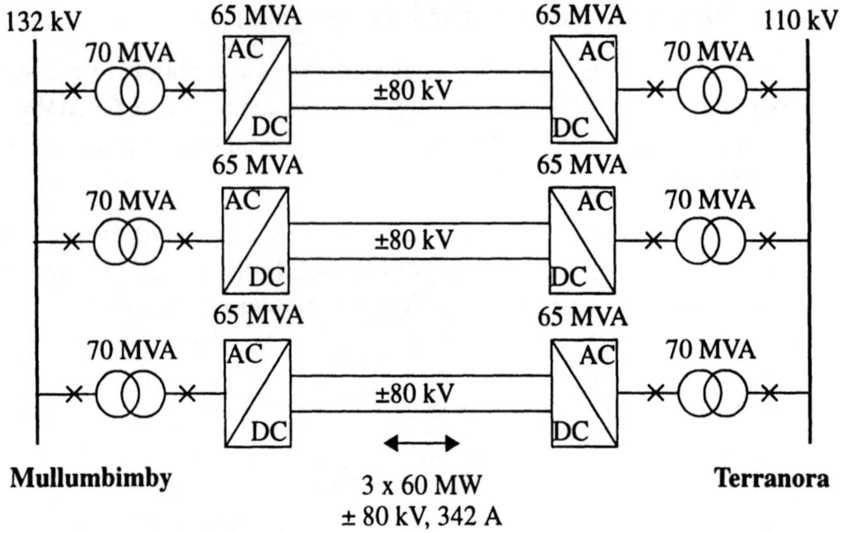

System Data:

Transformers	
Rated power	70 MVA
Turns Ratios	132/78.5 and 110/78.5
HVDC Cables	
Cross section	1x630 mm^2, Aluminium
Outer diameter	52 mm
Weight	3.3 kg/m
Length	6x59 km
AC Filters (per converter)	
39 th harmonic	1 branch
78 th harmonic	1 branch
IGBT Valves	
Valve type	2-level
Cooling system	Water
IGBT type	2.5 kV/500A

Figure 8-14: Directlink project in Australia

Figure 8-15: One of the Directlink converter stations [ABB]

8.7.3 Control System

The VSC terminals act independently to provide ancillary services (such as VAR support) in the networks to which Directlink connects. The flow of energy over HVDC Light facilities can be precisely defined and controlled, thereby meeting NECA's Safe Harbour Provisions. The ability to control power flow over the facility also means that the capacity rights required for fully commercial networks service are readily defined. The control system used is a modern digital microprocessor based system which is user-friendly and includes control features such as:

- Active and reactive power control,
- DC and ac voltage control, and
- Power quality control.

8.8 CONCLUDING REMARKS

The past decade has seen the application of dc transmission technology to a new niche market due to the advances in fast high-power, high voltage switches and dc cable technology. The global tendencies and interests in reducing green house gas emissions has led to developments in renewable energy sources such as solar and wind power generation. This market is still growing in importance. The combined impact of market-pull and technology-push will lead to further developments in coming years.

8.9 ACKNOWLEDGEMENT

Much of the information presented in this chapter has been obtained from technical brochures and information supplied by ABB and Siemens.

8.10 REFERENCES

[1]. H.Jiang and A.Ekstrom, "Multi-terminal HVDC systems in urban areas of large cities", IEEE Trans. on Power Delivery, Vol.13, No.4, Oct. 1998. pp 1278-1284

[2]. L.Stendius and K.Eriksson, "HVDC Light - An excellent tool for city centre infeed", PowerGen Conference, Singapore, Sept. 1999.

[3]. F.Schettler, H.Huang and N.Christl, "HVDC transmission systems using voltage sourced converters - design and applications", IEEE Summer Power Meeting, July 2000, Vol.2, pp 715-720.

[4]. K.Eriksson, T.Jonsson and O.Tollerz, "Small scale transmission to ac networks with HVDC Light", 12th CEPSI Conference in Pattaya, Thailand, Nov. 1998.

[5]. G.Asplund, K.Eriksson and O.Tollerz, "Land and sea Cable interconnections with HVDC Light", CEPSI 2000 Conference in Manila, Phillipines, Oct. 23-27, 2000.

Chapter 9

Active Filters

9.1 INTRODUCTION

Power conversion by virtue of its basic role produces harmonics due to the slicing of either voltages or currents. To a large extent the pollution in the utility supply and the deterioration of the power quality has been generated or created by non-linear converters. It is therefore ironic that power converters should now be used to clean up the pollution that they helped to create in the first place.

In a utility system, it is desirable to prevent harmonic currents (which result in EMI and resonance problems) and limit reactive power flows (which result in transmission losses).

Traditionally, shunt passive filters, comprised of tuned LC elements and capacitor banks, were used to filter the harmonics and to compensate for reactive current due to non-linear loads. However, in practical applications these methods have many disadvantages.

In the last two decades, considerable progress has been made in the field of Active Filters (AFs). AFs are inverter circuits, comprising of active devices i.e. semiconductor switches that can be controlled so as to act as harmonic current or voltage generators. Different topologies and control techniques have been proposed for their implementation. AFs are superior to passive filters in terms of filtering characteristics and improve the system stability by removing resonance related problems.

With remarkable progress in the speed and capacity of semiconductor power switching devices (i.e. GTO thyristors and IGBTs), AFs composed of voltage or current source inverters are being put to practical use, because they have the ability to overcome the disadvantages inherent in passive filters. Moreover, AFs act as "harmonic eliminators" rather than "harmonic attenuators", thus improving filtering characteristics.

In 1976, L. Gyugyi and E. C. Strycula [2] presented a family of shunt and series AFs, and established the concept of AFs consisting of PWM inverters using power transistors. However, these AFs could not be realized in real power systems because high-power high-speed switching devices were unavailable in the 1970's.

The work laid down the fundamental principles in the application of AFs. Later research evolves around the same fundamental concepts, but advancing towards newer control techniques for generation of switching signals. With the remarkable developments in switching speed and capacity of power semiconductor devices in the 1990's, AFs consisting of PWM inverters have been put to practical applications in real power systems.

Earlier work on AFs discussed compensation characteristics only for steady state conditions. The calculation circuit for the compensating current references was simple, and only ideal compensation characteristics could be attained. However, in transient states such as those caused by fluctuating loads, the design of compensating current calculation circuit becomes difficult.

In 1984, H. Akagi *et al.* [3] introduced a new concept of instantaneous reactive power. It dealt with 3 phase voltages and currents considering their distortion content. The instantaneous voltages and currents were represented as instantaneous space vectors on the a-b-c coordinate system, with the a,b,c vectors fixed on the same plane, apart from each other by 2*pi/3. These space vectors were then transformed along the alpha-beta orthogonal coordinate system. The instantaneous real and reactive power defined along the alpha-beta coordinates have a dc and ac component. The ac component pertains to the reactive and harmonic current of the load. The instantaneous active and reactive power were computed on-line and the ac component was extracted with a suitable filter. The design of the extraction filter has a significant effect on the compensation characteristics of the active filter. With the choice of cut-off frequency, the AF can be made to compensate for only the harmonic current or the reactive current or both. However, the theory was conceptually limited to three phase systems without zero-sequence currents.

A generalized instantaneous reactive power theory which is valid for sinusoidal or non-sinusoidal, balanced or unbalanced three phase power systems with or without zero-sequence currents was later proposed by F. Z. Peng and J. S. Lai [4]. AFs controlled on the basis of instantaneous reactive power theory provided good compensation characteristics in steady state as well as transient states.

At the same time, the following problems of AFs were pointed out:

- It was difficult to realize high power PWM inverters with rapid current response and low loss for use as a main circuit of AFs.
- The initial cost was high as compared with that of passive filters, and AFs were inferior in efficiency compared to passive filters.
- Injected currents by shunt AFs could flow into shunt passive filters and capacitors connected to the power system.

Therefore attention was paid to combined or hybrid systems of AFs and shunt passive filters.

In [5], the authors presented a novel compensation scheme using a shunt AF along with a conventional shunt passive filter. By sharing their roles so that the AF absorbs lower order harmonic currents and the passive filter absorbs higher order ones, the AF can fulfil its function with relatively small capacity, which brings an economical system. A lead function was introduced in its controller which brought about an interesting characteristic that the AF can act as a damping device in a parallel resonance circuit formed by the passive filter and the power supply system. This made it possible to realize an ideal harmonic filter with no amplification due to the parallel resonance over a whole range of harmonic orders. Experimentally the technique was verified on a cycloconverter load (2 sets of 2800 kW), being compensated with a 900 kVA AF and a 6600 kVA passive filter.

In [8], the authors proposed a combined system of a shunt passive filter and a small rated series AF. The combined system gave better filtering characteristics and lower initial and running costs. The technique was verified on a 20 kVA 3-ph, thyristor load, compensated by a shunt passive filter of 10 kVA and a series AF of 0.45 kVA. The function of the series AF is not to directly compensate for the harmonics of the rectifier, but to improve the filtering characteristics of the shunt passive filter and to solve the resonance related problems of a passive filter used alone.

It was shown that the series AF acts as a "damping resistance" which could eliminate the parallel resonance between the shunt passive filter and the source impedance, and also acts as a "blocking resistance" which could prevent the harmonic current produced by the source harmonic voltage from flowing into the shunt passive filter.

But for moderate power ratings, AFs function well independently with small power rated filters connected to eliminate switching voltages and currents. Different control techniques like PWM carrier based error sawtooth control [5,8,12,15], hysteresis based dead band control [3,7,10,13], magnetic flux compensation have been proposed and experimentally verified.

In [9], an AF for a 3-phase controlled rectifier system was presented. The AF comprised of a 3-phase voltage source inverter (VSI) with a dc bus capacitor source, connected in shunt to the ac source. The converter used two control loops. The outer control loop regulated the voltage across the dc bus capacitor, and generated reference phase AF currents. The inner control loop used sliding mode control to shape the AF currents in accordance with the reference currents.

In this work, the following assumptions were made:

- An infinite (strong) ac system was assumed. Since source impedance was not considered, the problem of voltage distortion at the terminal end does not occur.
- The line currents supplied by the source are in phase with the source voltages.
- An isolation transformer served as the connection impedance between the AF and the ac system.
- The sliding mode control principle was used to check the state of the system at a constant decision frequency, and to generate appropriate gating signals for the AF switches.

The sliding mode control principle used is easy to implement, and sets an upper limit to the switching frequency. The authors gave a good analysis of the effect of decision frequency, dc bus voltage, and variation of load on the THD of source current, power factor and efficiency of the system. The AF was able to supply load harmonics and limit the source current THD to 12-13% at full load.

In [10], a different control scheme was used to estimate the reactive and harmonic content of the non-linear load. The system here too, comprised of a shunt connected 3-phase voltage source active filter.

But here also, the effect of source impedance was not considered. The non-linear load considered was a diode-rectifier feeding a capacitive load. The control scheme measured the active power requirements of the load on-line, and generated reference currents for the AF switches. The switching signals for the AF switches were genèrated by a hysteresis based controller. The scheme provided excellent performance in terms of limiting the THD of source current to less than 5% at full load, along with a good efficiency.

Although hysteresis controllers provide excellent current tracking, they inherently rely on having a variable switching frequency. However, a proper choice of the hysteresis band and the connection impedances limits the average switching frequency to within the switching constraints. The work also highlighted the fast response of the AF to transient and load variations.

9.2 DC FILTERS

For a discussion on dc active filters, see chapter 15.

9.3 AC FILTERS

The AF control scheme presented here provides on-line computation of the active power component of the load and the internal AF losses. Therefore, the ac source current comprises of two components:

- Fundamental-frequency, active-power component of load current, and
- Fundamental-frequency, active-power component to maintain the dc bus capacitor voltage constant.

This computed source current is subtracted from the sensed load current to generate AF reference currents. A hysteresis-based current controller generates switching signals for the AF to follow the reference currents within specified band-limits [20].

The AF thus, acts as a local source supplying the load harmonics and reactive power. This improves the utility supply system power factor as the ac source provides only active fundamental frequency currents.

9.3.1 Test System

The system configuration of the AF is shown in Figure 9-1. A weak 3-phase ac source with 20% source impedance powers the system bus. An isolation transformer with 20% base impedance isolates the 6 pulse thyristor bridge rectifier from the system bus. The rectifier firing angle is controlled by a PI current controller (with gains K_p and K_i). The non-linear load consists of a series *R-L* load.

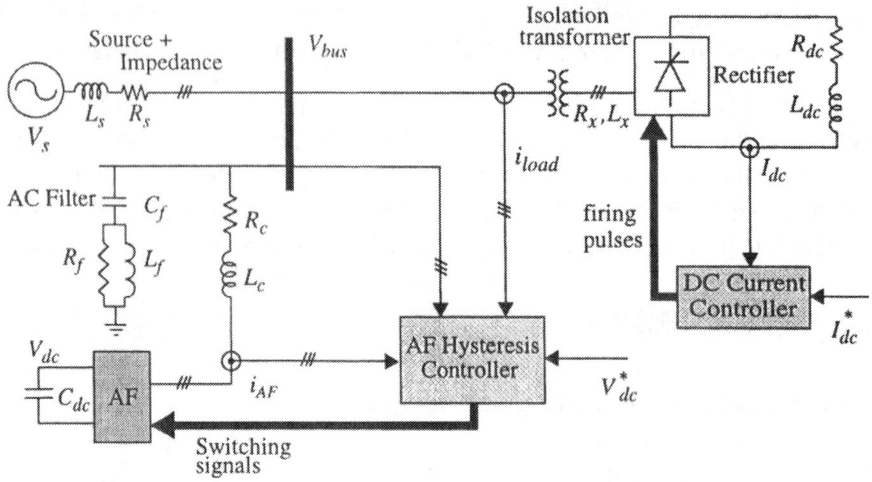

Figure 9-1: Test system

The AF used is a 6-switch VSI connected in parallel at the system bus. A dc capacitor C_{dc} acts as its energy storage/source, and provides the reactive and harmonic powers for the non-linear load and a small amount of active power to meet internal losses occurring in the form of switching and dielectric losses. The dc capacitor voltage V_{dc} is maintained constant, and fluctuates only during transients. The AF is connected to the system bus with a series inductive impedance (R_c, L_c) to shape the AF current. The replication capability, within hysteresis band-limits, of the AF currents to its reference currents is controlled by this series impedance.

As the AF currents may differ from their reference values within the hysteresis band-limits, high frequency (hf) components are introduced in the source currents and cause deviation in the system bus voltage. To reduce this hf current ripple, a high-pass filter (tuned to 12th harmonic) is connected at the system bus. This ensures that the source impedance carries near fundamental frequency current, and that the system bus voltage has an acceptable ripple component.

9.3.2 Control Philosophy

The objective of the control philosophy is to maintain a unity power factor (UPF) supply current by feeding an active-power current component to the:

- Non-linear load, and
- Active filter.

The reference value of the current component to the load I^*_{smp} is computed using the sensed average load power P_{ave}. The sensed load currents (i_{La}, i_{Lb}, i_{Lc}) and sensed bus voltages (v_a, v_b, v_c) are used to derive the instantaneous power p_L, as shown by Tanaka and Akagi [5]. From this value, the average load power P_{ave} is derived by averaging over 1/6 cycle period T_x.

The reference value of the peak current component to the AF I^*_{smd} is computed using the value of the capacitor C_{dc}, the average voltage $V_{dc\text{-}ave}$ and the desired capacitor voltage V^*_{dc}.

The total (peak) reference source current I^*_{sm} is computed as the sum of these two components i.e.

$$I^*_{sm} = I^*_{smp} + I^*_{smd} \tag{9-1}$$

The 3 phase instantaneous reference source currents (i^*_{sa}, i^*_{sb}, i^*_{sc}) are computed using this peak value I^*_{sm} and unit current templates (u_{sa}, u_{sb}, u_{sc}) derived from sensed bus voltages (v_a, v_b, v_c) and the peak value V_{pk} of the bus voltage V_{bus}.

The desired references of the AF current (i^*_{ca}, i^*_{cb}, i^*_{cc}) are computed by taking the difference between the 3 phase instantaneous reference source currents (i^*_{sa}, i^*_{sb}, i^*_{sc}) and sensed load currents (i_{La}, i_{Lb}, i_{Lc}), i.e.

$$\left.\begin{aligned} i^*_{ca} &= i^*_{sa} - i_{La} \\ i^*_{cb} &= i^*_{sb} - i_{Lb} \\ i^*_{cc} &= i^*_{sc} - i_{Lc} \end{aligned}\right\} \tag{9-2}$$

The hysteresis rule base, similar to techniques used by Ziogas [6] and Kawamura and Hoft [7], is then employed over the reference and sensed AF currents to obtain the switching signals for the AF.

The control philosophy is explained with the help of the schematic shown in Figure 9-2. Three major blocks are considered in the controller as follows:

9.3.2.1 Block 1: Derivation of Component I^*_{smp},

The first component of the source current is I^*_{smp} the component due to the load current. The sub-blocks in this unit consist of the following:

(A) Compute Peak Bus Voltage. This sub-block is used to derive the peak bus voltage from the sensed 3 phase voltages. Under ideal conditions, the 3 phase voltages are expressed as:

$$\left.\begin{aligned} v_a &= V_{pk.} * sin(wt) \\ v_b &= V_{pk} * sin(wt + 2*Pi/3) \\ v_c &= V_{pk} * sin(wt - 2*Pi/3) \end{aligned}\right\} \quad (9\text{-}3)$$

(B) Compute Instantaneous Load Power. The instantaneous load power p_L is obtained by:

$$p_L = v_a * i_{La} + v_b * i_{Lb} + v_c * i_L \quad (9\text{-}4)$$

where:

(v_a, v_b, v_c) - are the sensed 3-phase load voltages, and

(i_{La}, i_{Lb}, i_{Lc}) - are the sensed 3-phase load currents.

(C) Compute (peak) Load Current Reference. The sensed average load power over one sixth of a cycle period is given by:

$$P_{ave} = 1.5 * V_{pk} * I^*_{smp} \quad (9\text{-}5)$$

where:

V_{pk} is the sensed peak of the bus voltages, and

I^*_{smp} is the desired (peak) reference of the load current.

Since the values of P_{ave}, and V_{pk} can be sensed, the value of I^*_{smp} can be obtained from the above equation.

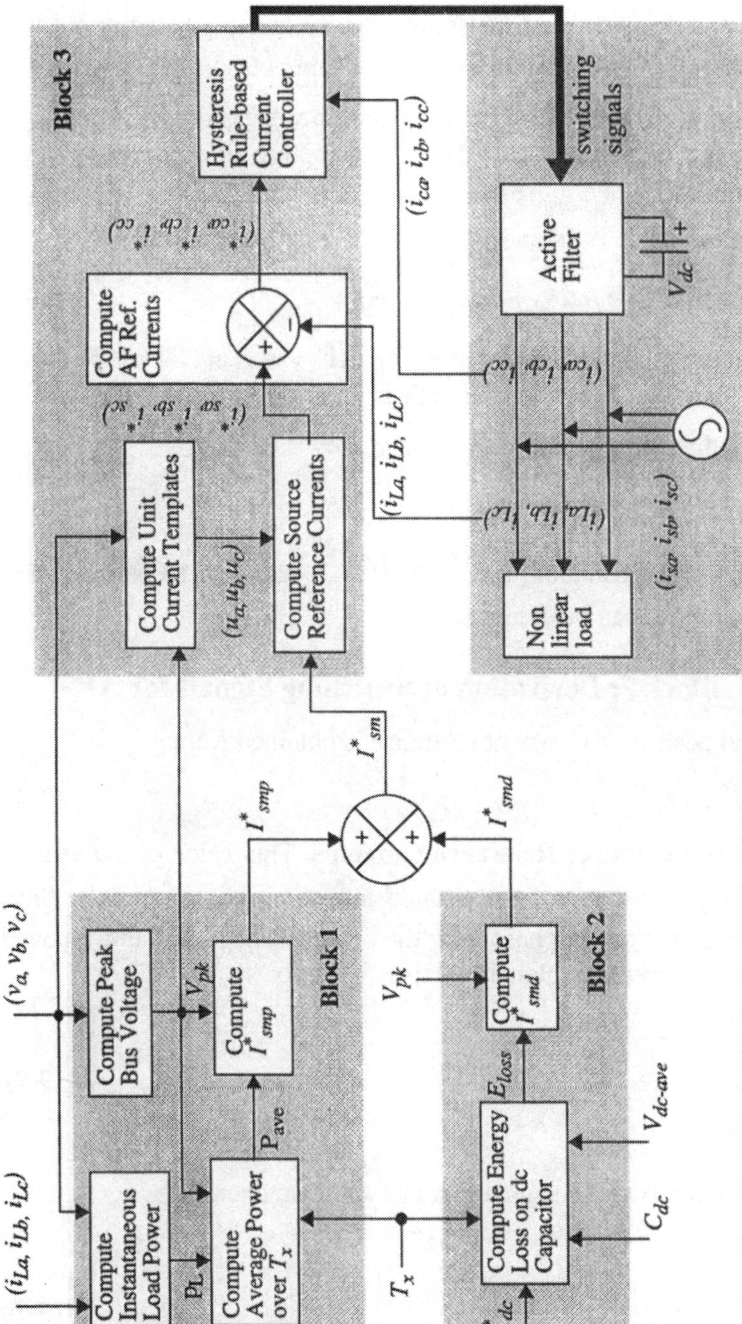

Figure 9-2: Control philosophy [20]

9.3.2.2 Block 2: Derivation of Component I^*_{smd},

The second component of the source current is I^*_{smd} the component due to the AF current. The two sub-blocks in this unit consist of:

(A) Compute Energy Loss on dc Capacitor. A desired reference dc capacitor voltage is assumed as V^*_{dc}. The sampling of the average actual dc capacitor voltage $v_{dc\text{-}ave}$ is done every one-sixth of a cycle. The energy loss over the period T_x is given by:

$$E_{loss} = 0.5 * C_{dc} * [(V^*_{dc})^2 - (V_{dc\text{-}ave})^2] \qquad (9\text{-}6)$$

(B) Compute (peak) Reference for AF Current. The AF draws this energy loss E_{loss} from the source through a UPF current with a peak value of I^*_{smd}; this energy relationship is given by:

$$E_{loss} = 1.5 * V_{pk} * I^*_{smd} * T_x \qquad (9\text{-}7)$$

From the above equation, the value of I^*_{smd} can now be derived, since the other quantities can be computed.

9.3.2.3 Block 3: Derivation of Switching Signals for AF,

The total peak source current reference is obtained from:

$$I^*_{sm} = I^*_{smp} + I^*_{smd} \qquad (9\text{-}8)$$

(A) Compute Source Reference Currents. This value of the peak source current I^*_{sm} must now be translated into three phase currents. Since the source currents are in phase with the source voltage (i.e. unity power factor), a unit current template is obtained as follows:

$$\left.\begin{aligned} u_a &= v_a/V_{pk} \\ u_b &= v_b/V_{pk} \\ u_c &= v_c/V_{pk} \end{aligned}\right\} \qquad (9\text{-}9)$$

The reference 3 phase source currents are then given as:

$$\left.\begin{aligned} i^*_{sa} &= I^*_{sm} * u_a \\ i^*_{sb} &= I^*_{sm} * u_b \\ i^*_{sc} &= I^*_{sm} * u_c \end{aligned}\right\} \qquad (9\text{-}10)$$

(B) Compute AF Reference Currents. The 3 phase AF reference currents are derived as the difference of the reference source currents and the sensed load currents:

$$\left.\begin{aligned} i^*_{ca} &= i^*_{sa} - i_{La} \\ i^*_{cb} &= i^*_{sb} - i_{Lb} \\ i^*_{cc} &= i^*_{sc} - i_{Lc} \end{aligned}\right\} \quad (9\text{-}11)$$

(C) Hysteresis Rule-based Current Controller. The switching pattern for the AF is determined by the current controller, i.e. for phase a of the VSI):

If $i_{ca} < i^*_{ca}$ - hb, then the upper switch must be ON and the lower switch OFF.

If $i_{ca} > i^*_{ca}$ + hb, then the lower switch must be ON and the upper switch OFF.

The switching signals are generated similarly for the other phases, using the corresponding reference and measured currents and the hysteresis band.

The AF performance can be improved by narrowing the hysteresis band, but owing to the limitations of switching devices, a hysteresis band at ±5% of the current reference is considered an acceptable compromise.

The AF currents i_{ca}, i_{cb}, and i_{cc} are regulated to be in good agreement with the reference values i^*_{ca}, i^*_{cb}, and i^*_{cc}.

9.3.3 Test Results

9.3.3.1 Steady State Performance of the AF (Figure 9-3)

The results shown are the bus voltage, the current drawn by the rectifier load, the current supplied by the source and the AF current. The following observations can be made:

- The source current is in phase with the bus voltage i.e. unity power factor load is presented to ac bus,
- The ability of the AF to supply the load harmonic currents and reactive components is clearly evident,

- The ac bus voltage contains a hf ripple which can be reduced by a suitable choice of the passive filter, and
- Switching frequency of the AF is variable depending on the load current.

9.3.3.2 Transient Performance of the AF

Two tests were conducted for evaluation of the transient response of the AF and the non-linear load together:

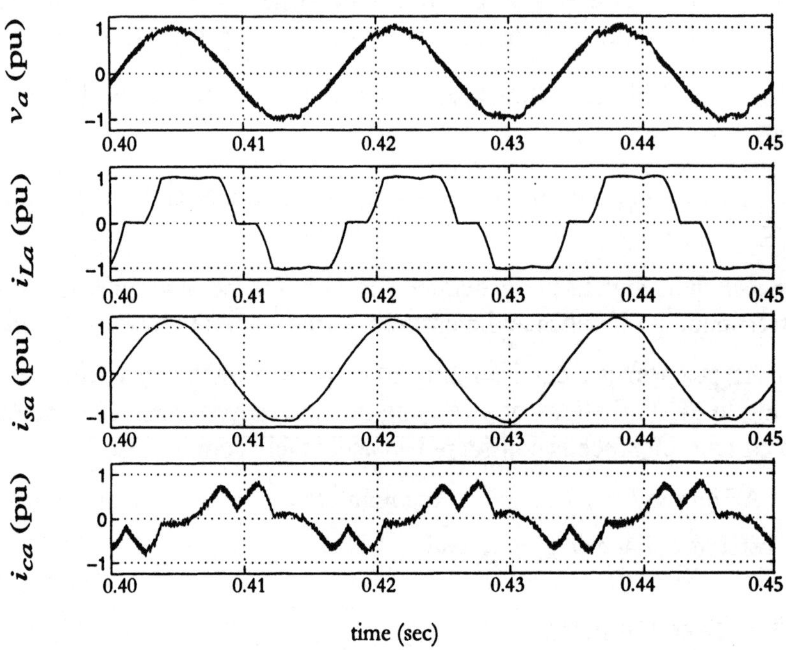

Figure 9-3: Steady state performance [20]

a) Step Change In I_{ref} Of The DC Controller (Figure 9-4)

A 50% step reduction in the current reference (I_{ref}) value of the dc load is applied at 0.25 s and removed at 0.35 s. Shown are the reference (I_{ref}) and measured dc current (I_{dc}), the bus voltage (v_a), the ac current into the non-linear load (i_{La}), the input source current (i_{sa}), the AF current (i_{ca}), and the dc capacitor voltage (V_{dc}). The dynamic responses of the dc controller and the AF are both rapid and stable.

b) Single Line To Ground (SLG) Fault At AC Bus (Figure 9-5)

During the 3-cycle SLG fault application, the dc current in the load is reduced and a second harmonic is injected into the load. The dc controller is forced to its alpha minimum limit of about 5 degs. from its nominal operating angle of about 15 degs. The dc system and the AF are able to recover within 1 cycle, although the harmonic content is reduced more slowly within about 3 cycles. During the fault period, the capacitor voltage V_{dc} is reduced slightly to accommodate the increased requirement for compensation demanded by the fault current. Although the decay of the dc component in the fault current is quite small due to limited damping losses, the ability of the AF to track the harmonic requirements of the load current are clearly evident.

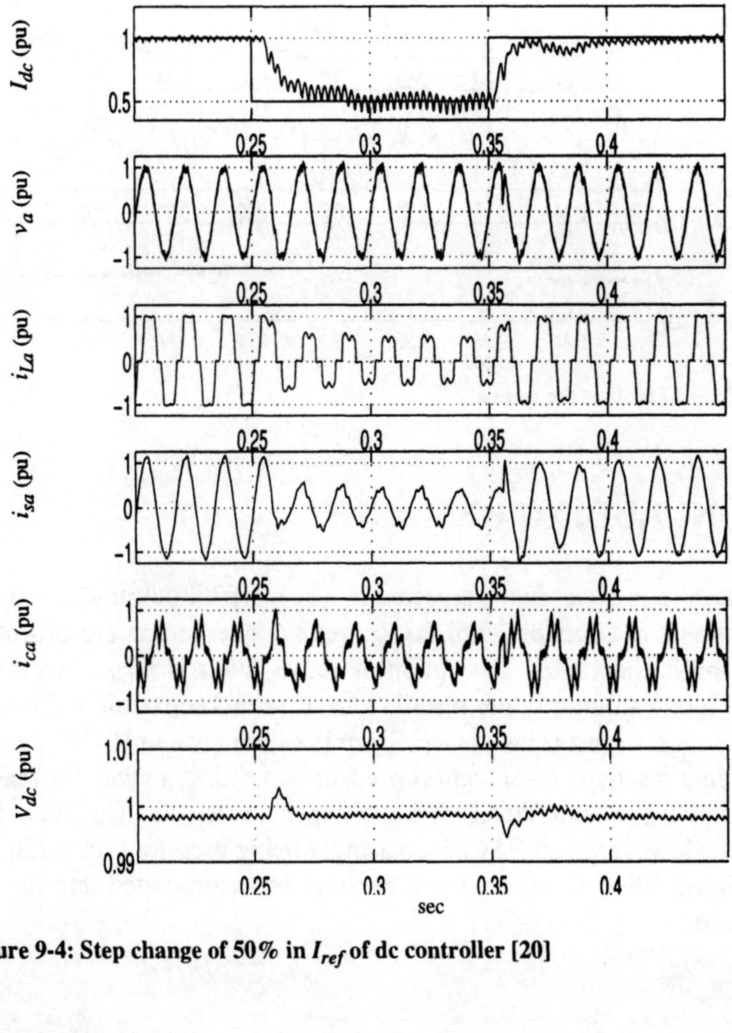

Figure 9-4: Step change of 50% in I_{ref} of dc controller [20]

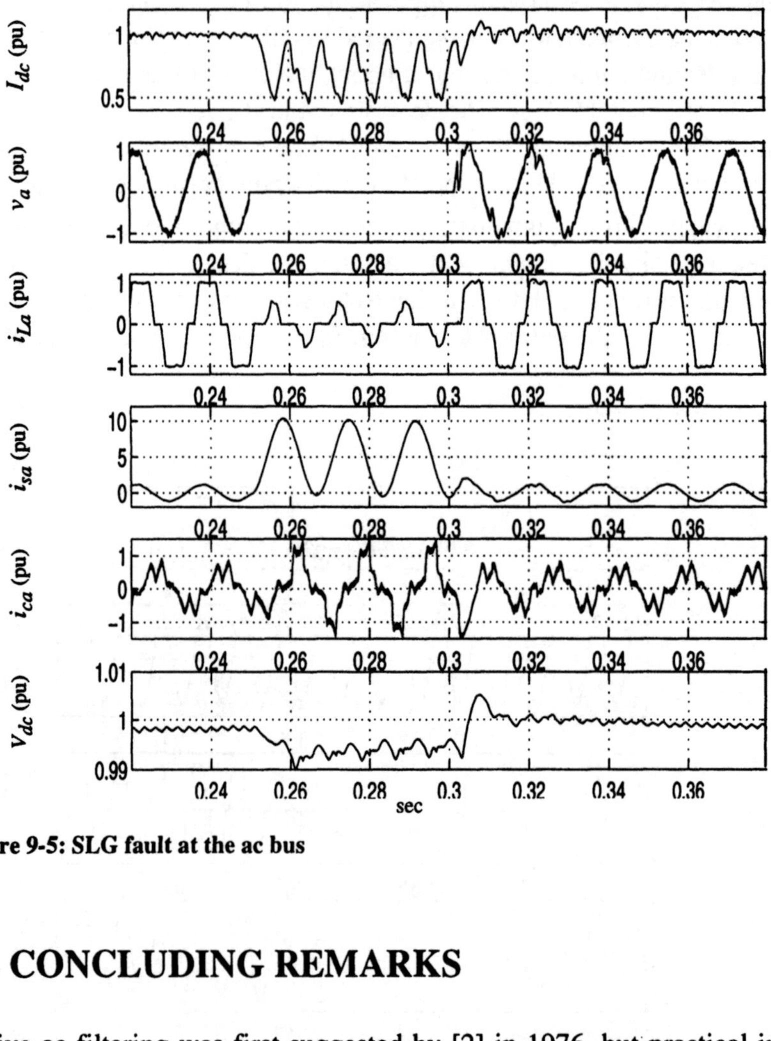

Figure 9-5: SLG fault at the ac bus

9.4 CONCLUDING REMARKS

Active ac filtering was first suggested by [2] in 1976, but practical implementation was not possible until fast power switches and micro-processors came along in the 1990s. The option of active filters is now a reality and will gain momentum in future installations as costs keep coming down and reliability improvements take place. Since line commutated HVDC converters consume reactive power, capacitors will be needed anyway for reactive power support; it will therefore be cheaper to use the ConTune filter approach. However, with VSCs increasingly being used for under 250 MW installations, active ac filtering will begin to be incorporated into the converter itself.

9.5 ACKNOWLEDGEMENT

Results from the thesis of Mr. R.Arora, a former student of mine at Concordia University, Montreal, have been used here for some cases.

9.6 REFERENCES

[1]. H.Sasaki & T.Machida, "A New Method to eliminate AC harmonic Currents by Magnetic Compensation Considerations on basic design", IEEE Trans. on Power Apparatus and Systems, 1971, Vol. PAS-90, No.5, pp. 2009-2019.

[2]. L.Gyugyi & E.C.Strycula, "Active AC Power Filters", IEEE-IAS Annual Meeting Record,1976, pp. 529-535.

[3]. H. Akagi, Y. Kanazawa & A. Nabae, "Instantaneous Reactive Power Compensators Comprising Switching Devices without Energy Storage Components", IEEE Trans. on IAS, Vol. IA-20, No.9, May/June 1984, pp. 625-630.

[4]. Fang Zheng Peng & Jih-Sheng Lai, "Generalized Instantaneous Reactive Power Theory for Three-Phase Power Systems", IEEE Trans. on Instrumentation and Measurement, Vol. 45, No.1, February 1996, pp. 293-297.

[5]. Masatoshi Takeda, Kazuo Ikeda & Yoshiharu Tominaga, "Harmonic Current Compensation with Active Filter", IEEE-IAS, 1987, pp. 808-815.

[6]. D.Sutanto, M.Bou-rabee, K.S.Tam & C.S.Chang, "Harmonic Filters for Industrial Power Systems", IEE Int. Conference on Advances in Power System Control, Operation and Management, Nov. 1991, Hong Kong, pp. 594-598.

[7]. G. Ledwich & P. Doulai, "Control Techniques of Active Power Filters", IEE International Conference on Advances in Power System Control, Operation and Management, November 1991, Hong Kong, pp. 582-587.

[8]. F.Zheng Peng, H. Akagi & A. Nabae, "A New Approach to Harmonic Compensation in Power Systems - A Combined system of Shunt Passive and Series Active Filters", IEEE-IAS, Vol. 26, No. 6, Nov/Dec 1990, pp. 983-990.

[9]. S. Saetieo, R. Devaraj & D. A. Torrey, "The Design and Implementation of a Three-Phase Active Power Filter Based on Sliding Mode Control", IEEE Trans. on Industrial Applications, Vol. 31, No. 5, Sept/Oct 1995, pp. 993-1000.

[10]. Bhim Singh, Kamal Al-Haddad & Ambrish Chandra, "A New Control Approach to 3-phase Active Filter for Harmonics and Reactive Power Compensation", IEEE Trans. on Power Systems, April 1997.

[11]. Toshihiko Tanaka & Hirofumi Akagi, "A New Method of Harmonic Power

Detection Based on the Instantaneous Active Power in Three-Phase Circuits", IEEE Trans. on Power Delivery, Vol. 10, No.4, October 1995, pp.1737-1742.

[12]. Phoivos D. Ziogas, "The Delta Modulation Technique in Static PWM Inverters", IEEE Trans. on Industry Applications, Vol. IA-17, No.2, March/April 1981, pp. 199-204.

[13]. Atsuo Kawamura & Richard Hoft, "Instantaneous Feedback Controlled PWM Inverter with Adaptive Hysteresis", IEEE Trans. on Industry Applications, Vol IA-20, No.4, July/August 1984, pp. 769-775.

[14]. Hirofumi Akagi, "Trends in Active Power Line Conditioners", Proceedings of IECON, San Diego, October 92, pp. 19-24.

[15]. Fang Zheng Peng, Hirofumi Akagi & Akira Nabae, "A Novel Harmonic Power Filter", IEEE-PESC Record, April 1988, pp. 1151-1158.

[16]. Hideaki Fujita & Hirofumi Akagi, "A Practical Approach to Harmonic Compensation in Power Systems - Series Connection of Passive and Active Filters", IEEE Conference Record IAS, Seattle, October 1990, Vol. 2, pp. 1107-1112.

[17]. P.F.Wojciak & David A.Torrey, "The design and implementation of active filter systems using variable structure system concepts", IEEE-IAS Annual Meeting Conf. Record, pp 850-857, 1992.

[18]. L. Gyugyi, T. R. Rietman, A. E. Edris, "The Unified Power Flow Controller: A New Approach to Power Transmission Control", IEEE Transactions on Power Delivery, Vol. 10, No.2, April 1995, pp. 1085-1097.

[19]. L. Gyugyi, "Unified power-flow control concept for flexible AC transmission systems", IEE Proceedings-C, Vol. 139, No. 4, July 1992, pp. 323-331.

[20]. V.K.Sood and R.Arora, "Development of an EMTP based model of an active filter for distribution system studies" IEE Int. Conf. on Power Electronics and Variable Speed Drives, Publication # 456, PEVD'98, London, UK. 21 Sept. 98, pp 181-186.

Chapter 10

Typical Disturbances in HVDC Systems

10.1 INTRODUCTION

Some of the typical disturbances that a dc system can be subjected to are listed below:

- Step changes in I_o and gamma order,
- AC faults on Rectifier Bus (1-Phase or 3-Phase),
- AC faults on Inverter Bus (1-Phase or 3-Phase),
- Loss of Firing Pulses (i.e. commutation failures),
- DC Line Faults,
- Power Reversal, and
- Block - Deblock of Rectifier.

It is imperative that the dc controllers are able to operate stably and adequately in the presence of such system disturbances. During the testing and verification period of dc controllers for an actual plant, such tests are applied in a simulator (physical or numerical) environment to test and optimize the dc controllers. Typical features and behavior patterns are presented in this chapter to assist readers to evaluate the performances and point out features to observe.

The results have been selected from an EMTP (version 3) study with a 6-pulse modified version of the CIGRE HVDC Benchmark model [1]. The reason for using a 6-pulse system is purely for the conservation of simulation time and data. However, this should not detract from the essential features about the controller and system inter-actions that are highlighted here.

First, the model of the CIGRE Benchmark is presented. Second, details of the control system used are presented. Finally some typical results with comments highlighting particular features are presented.

10.2 CIGRE BENCHMARK MODEL FOR HVDC CONTROL STUDIES

The CIGRE benchmark model for HVDC control studies was first proposed by J.Ainsworth. It has evolved to become the *defacto* standard HVDC benchmark for control studies and provides a useful model as a result of its many special characteristics [1]. In particular some characteristics were particularly selected to make this benchmark a difficult system to operate:

- The dc system was close to resonance at the fundamental frequency of 50 Hz,
- The ac systems were selected to be weak, and were close to resonance at the second harmonic.

In practice, a 12-pulse, 50 Hz converter system is specified in the CIGRE benchmark. To facilitate explanation of fundamental principles, reduce computer simulation time and minimize memory management, only an equivalent 6-pulse system is modeled here. The power system model with the components modeled is shown in Figure 10-1. Consequent to the decision to use a simplified model, it becomes necessary to compensate for the 6-pulse harmonics generated. Therefore, single-tuned 5th and 7th harmonic ac filters are necessary and incorporated into the system [2,3]. Design parameters used are provided here. The capacitor banks were suitably modified to account for the reactive power supplied from the 5th and 7th harmonic ac filters.

One necessary detail not supplied in the original Benchmark model are the details of the valve snubber circuits (Figure 10-2). These are important details for a good simulation model, and the parameters selected are provided here. It is noteworthy that the impact of the snubber resistance is much more pronounced than the snubber capacitance. The snubber values are based upon a fictitious equivalent valve characteristic, rather than on an individual switch.

Also due to the 6-pulse model simulated here, only a simplified dc filter is used, since the dc cable with its large capacitance acts as a good filter. This is justifiable since the objective here was to look at controller dynamics rather than dc system transients. Therefore, the dc line was modelled simply with only one T-section. Details of the dc filters used are provided in Figure 10-3. An identical filter is used at the rectifier as well as the inverter ends.

No ac machine dynamics are modelled in the ac systems. The equivalent fixed frequency ac systems use a 2L-R representation with the correct damping angle at fundamental and third harmonic frequencies.

The converter transformer models used a suitable leakage impedance and saturation characteristics. The leakage was equally distributed on the primary and secondary sides of the converter transformers. Only the star-star converter transformer configuration was used.

The benchmark was designed to enable manufacturers/utilities to compare the performance of their own HVDC controls against other units. No details of the controllers were provided and designers were free to utilise proprietary designs.

Figure 10-1: HVDC benchmark model [2]

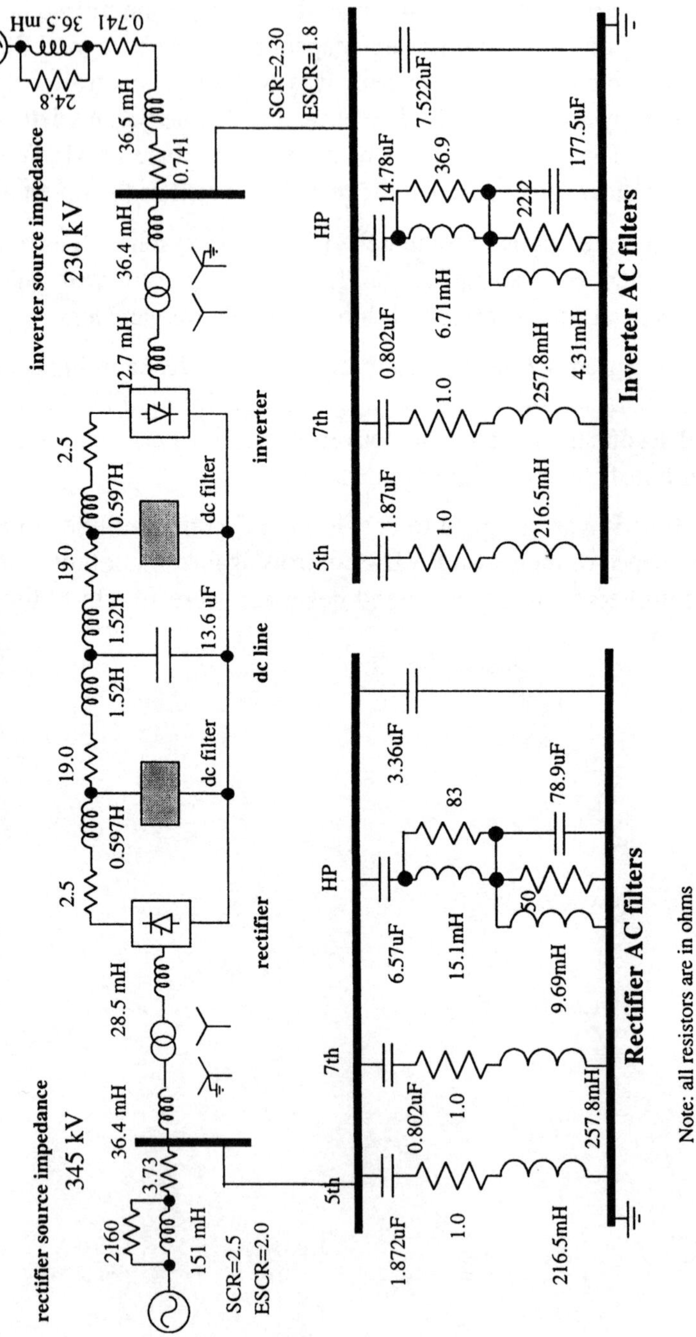

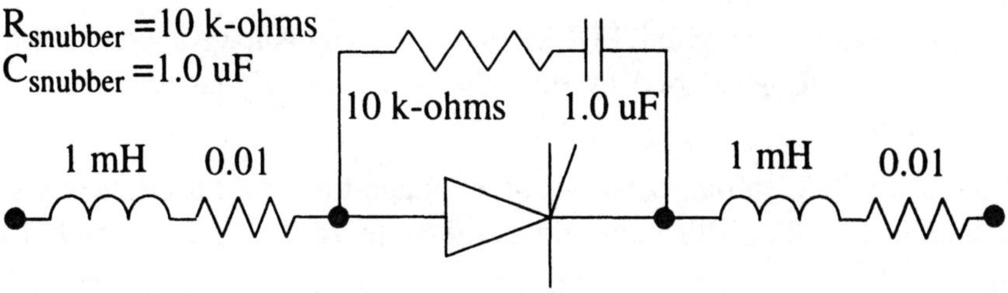

Figure 10-2: Converter switch model [2]

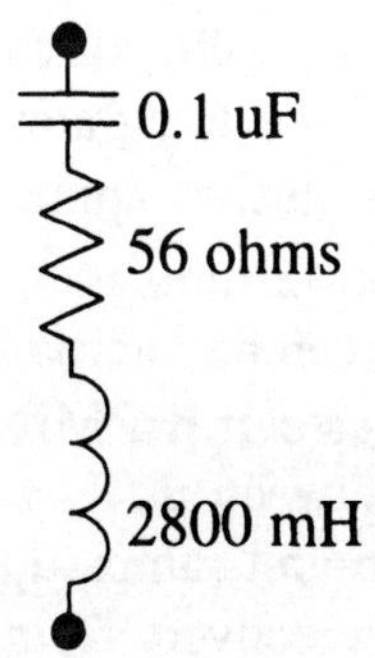

Figure 10-3: DC filters used [2]

10.3 DETAILS OF CONTROL SYSTEMS USED

10.3.1 Rectifier Control Unit

The rectifier control system is shown in Figure 10-4. The limited current IOLIM reference is generated using the Voltage Dependent Current Limit (VDCL) unit.

The VDCL unit provides IOSTAT and IODYN current references which are generated from the Static and Dynamic sub-units. These units provide current reference values during steady- and transient-state conditions respectively. The measured dc voltage is passed through:

(a) A low-pass filter VDCFIL (cut-off frequency = 30 Hz) to eliminate higher order harmonics, and
(b) A pulse-shaping unit PULSP to remove any voltage spikes. The output PULSP is used by both the Static and Dynamic VDCL sub-units.

The static VDCL sub-unit consists of a voltage-to-current transducer with a gain of 1.25 (Note: This gain effects the slope of the V_d-I_d characteristic, and may be adjusted to provide different characteristics) and is limited between minimum IMIN and maximum IMAX current limits (Figure 10-5). The limited output IDCLIM is fed to a MINIMUM-selector whose other input is the current reference IREF. The output of the MINIMUM-selector is IOSTAT.

The dynamic VDCL sub-unit consists of an integrator (Time constant 50 ms) which dictates the rise time of the dc current during the fault recovery period; the integrator input is a bi-directional signal shaped from the dc voltage. The RAMP output of the integrator is passed through a MINIMUM-selector which has the pulse-shaping circuit and I_{ref} as its other inputs. The MINIMUM-selector ensures that, in the worst case, the output of these blocks is IREF, which can be internally selected. IMIN is added to the output of the MINIMUM-selector IRRMIN through a summer. The output of the summer is limited between IMIN and IREF. This forces the output of the dynamic VDCL sub-unit ramps up from IMIN to IREF during system start-up, and also during recovery from faults such as a dc line fault.

The measured rectifier dc current IDCR is compared to the limited current reference IOLIM from the VDCL unit and an error signal is generated. This error is fed to a PI controller (limited within alpha-min (AMIN) = 5 degrees to alpha-max (AMAX) = 145 degrees.) via a supplementary block, the other input of which is a delayed step input FORAND. This signal forces the PI controller to zero input for a period of 0.1s; the effect of this FORAND signal is to aid in the rapid initialization of the EMTP case. The output of the PI controller is an alpha order which is fed to a Ring Counter via the MAXIMUM-select block. The MAX block has an auxiliary signal ALPRET which is used during the recovery from a dc line fault. The ALPRET signal forces the alpha order fed to the Ring Counter to a value of alpha-max = 145 degrees and the converter into the inverter region of operation.

Further details of the trigger unit can be found in reference [2].

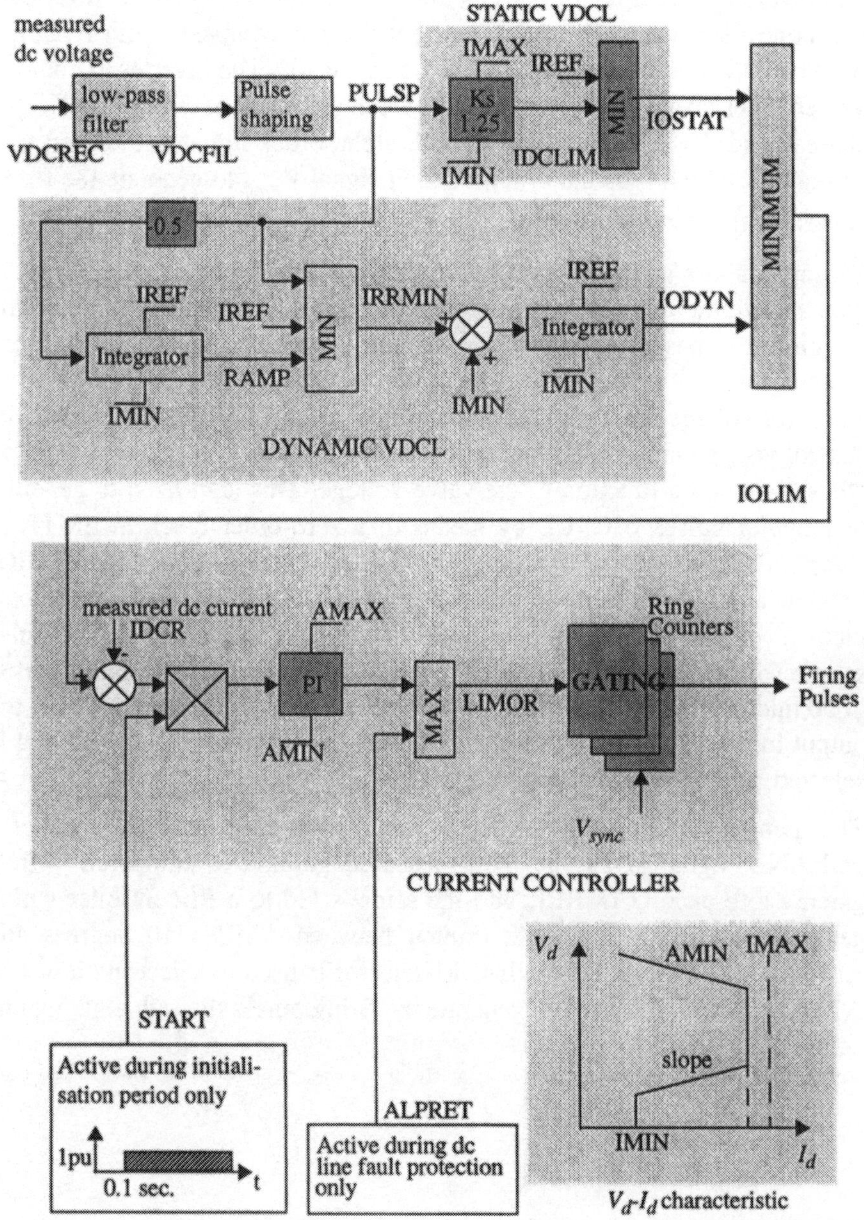

Figure 10-4: Rectifier VDCL and Current Controller Units

10.3.2 Inverter Control Unit

The inverter current controller (Figure 10-5) is similar to the rectifier current controller. The measured dc current IDCI is compared with IOLIM - ΔI, where the current margin is ΔI=0.1 pu typically. The inverter PI controller generates an alpha-order signal A1 which is limited between AMIN=110 degrees and AMAX=150 degrees. The alpha order signal is then fed to a gating unit which uses the synchronizing signal V_{sync} to generate the firing pulses for the inverter valves.

To implement a gamma controller, a measure of the gamma is required. A typical Inverter valve voltage is shown in Figure 10-5; the gamma measuring circuit is based on evaluating the time elapsed between the two zero crossings a and c of the valve voltage. The negative voltage a-b-c is used as a control voltage for a resettable integrator which integrates only when its control voltage is positive. The output of the resettable integrator is a ramp whose width is a function of the valve voltage. This is utilized to generate rectangular pulses BIGPUL by means of a zero-order block in EMTP. A sharp pulse (duration of one time-step) is derived at a rising edge of BIGPUL, which is used to reset the counter at the start of BIGPUL. The complement of BIGPUL is used to hold the count of the counter. The pulse source for the counter has a pulse-width of 25 micro-seconds and a period of 50 micro-seconds. Hence each count represents 0.9 degrees and to get the output in degrees, a scaling factor of 0.9 is used. The minimum gamma is selected, and used after some smoothing.

The gamma controller is similar to the current controller, except that a FORAND signal is absent. The measured gamma is compared with a gamma reference GAMREF, and the error is fed to a PI controller which generates an alpha order A2, limited between AMIN=110 degrees and AMAX=150 degrees. A MINIMUM-selector is used to select the lower of A1 and A2, which is used to generate the firing pulses through a gating circuit.

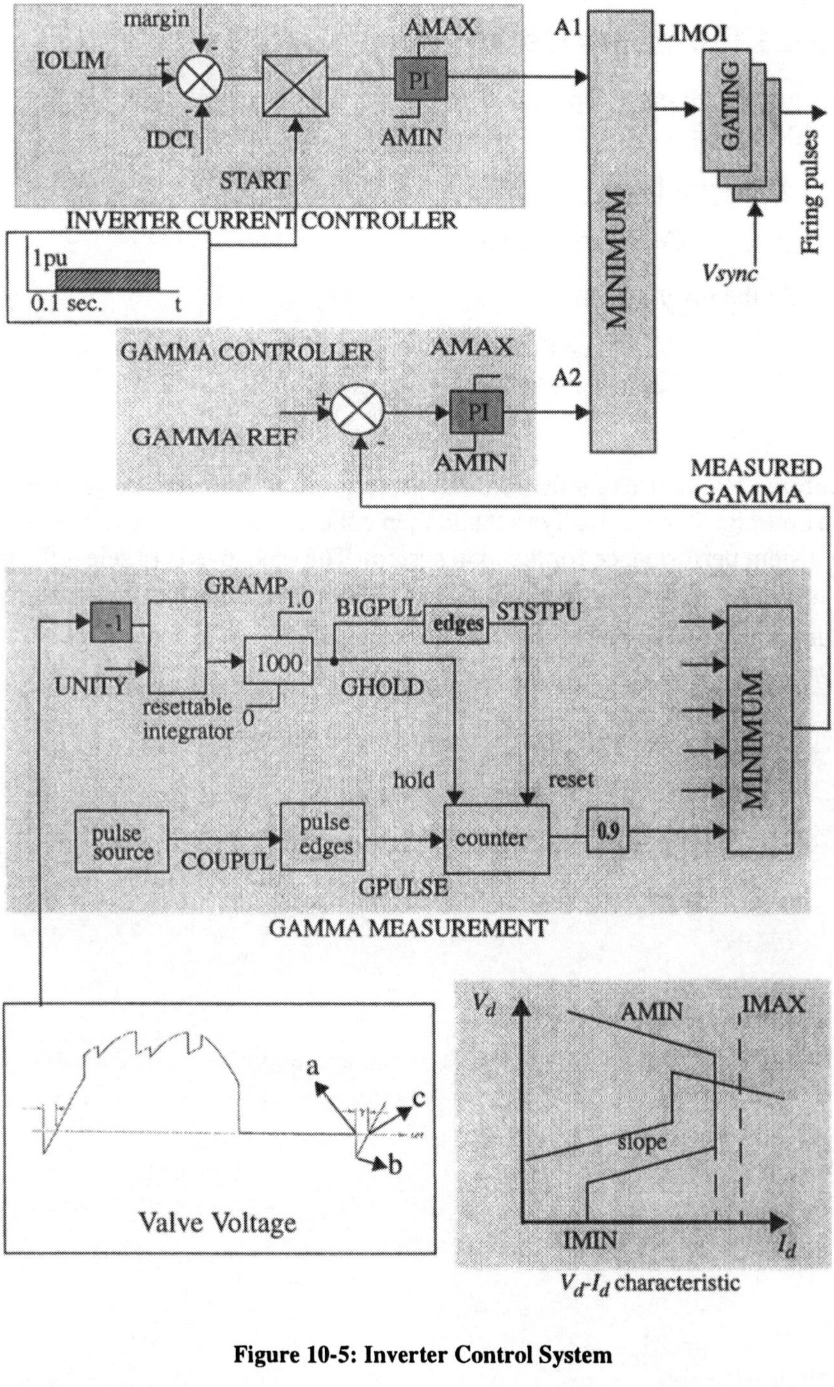

Figure 10-5: Inverter Control System

10.4 RESULTS

10.4.1 Controller Optimization Tests

During steady-state operation, two controllers are in operation for this HVDC system:

- At the rectifier:

  - Current controller,

- At the inverter:

  - Current controller, or
  - Gamma controller.

Before any other dynamic tests are performed, all three controllers need to be optimized, individually first and then collectively, to obtain a satisfactory transient performance for the total system. The objective is to select the gain parameters K_p and K_i for all three PI regulators. For this optimization procedure, the following three tests are recommended:

1. 10% step change in rectifier current reference,
2. 5% step change in inverter current reference, and
3. 2.5^o step change in inverter gamma reference.

10.4.1.1 10% Step Change In Rectifier Current Reference

To test the performance of the current controller at the rectifier end, a negative 0.1 pu step change is applied to its current reference of 1.0 pu (Figure 10-6). To effectively study the performance of the rectifier current controller without any influence from the inverter current controller, the current margin ΔI is temporarily set to 0.2 pu (i.e. greater than the 0.1 pu margin). The step change is applied at 1.3 s and the step change is effected in 100 ms. The response is both well controlled and stable. The results shows the dc current on the rectifier and the inverter side along with their super-imposed reference values (dashed curves). Minor differences in the harmonic content of the dc currents at the rectifier and inverter end are noticeable due to the dc line and dc filter characteristics. The dc voltages on the rectifier and the inverter sides are also shown. Essentially, the dc voltage remains almost constant since it is controlled by the gamma controller at the inverter end. Minor differences in the harmonic contents of the two dc voltage are observable. The rectifier alpha-order signal also shows the step change in the current being effected as the average value of alpha is transiently changed from about 14 degrees to about 22 degrees. Eventually the alpha order settles at about 16 degrees for a current at 0.9 pu. The inverter alpha-order signal (settled around 140 degrees) shows that the inverter current controller plays little or no role in controlling the inverter current and that the gamma controller controls the inverter voltage.

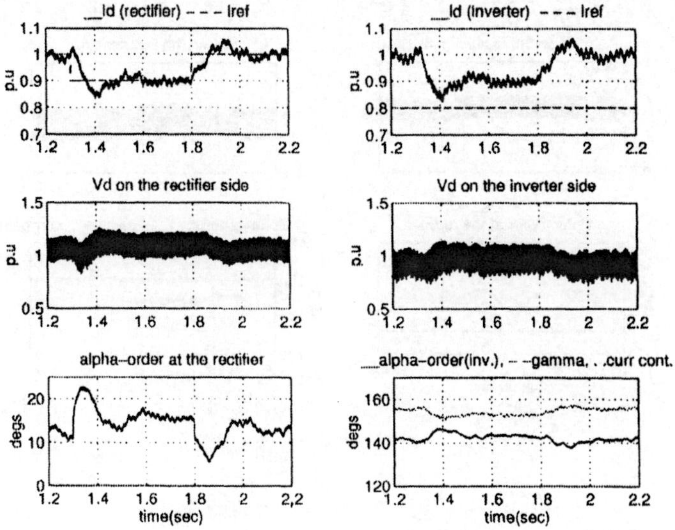

Figure 10-6: 10% Step change in rectifier current reference

10.4.1.2 5% Step Change In Inverter Current Reference

To test the performance of the inverter current controller, first the ac supply voltage at the rectifier-end is reduced by about 5% to force the rectifier controller to its alpha-min limit equal to about 9 degrees. This effectively permits study of the performance of the inverter current controller only, and prevents any interaction from the rectifier current controller. The effective dc current is reduced to 0.9 pu due to the current margin of 0.1 pu. Next, a 5% step reduction is then applied to its inverter current reference (Figure 10-7) at the 1.5 s point. The response to the step change is stable and well controlled and is effected in about 80 ms. The results show the dc current on the rectifier and the inverter sides along with their respective reference values (dashed lines). The dc voltages on the rectifier and the inverter sides are also shown and these remain essentially constant at about 1.1 pu due to the rectifier alpha-order signal being at alpha-min of 9 degrees. So the inverter current controller controls the dc current of the system.

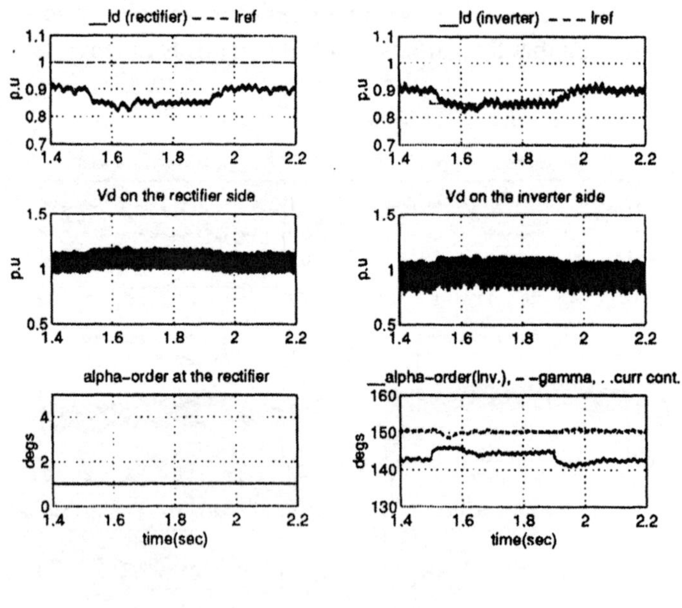

Figure 10-7: 5% Step change in inverter current reference

10.4.1.3 2.5° Step Change In Inverter Gamma Reference

To test the performance of the inverter gamma controller, a 2.5 degrees step change (increasing step at 1.6 s from an original gamma value of 20 degrees to 22.5 degrees is followed by a decreasing step at 1.9 s) is applied in the gamma reference (Figure 10-8). The response is reasonably well controlled and stable with a response time of 50 ms. (Note: It is also noted that some steady state oscillation in the dc current is present; this only serves to distract from the main test here, and is not relevant to the discussion at hand). The dc voltages at the rectifier and inverter dip in response to the increase of the gamma value. Consequently, a short-lived transient increase in the dc current is noticed. In response, the alpha order signal at the rectifier-end thus increases to maintain the dc system current constant.

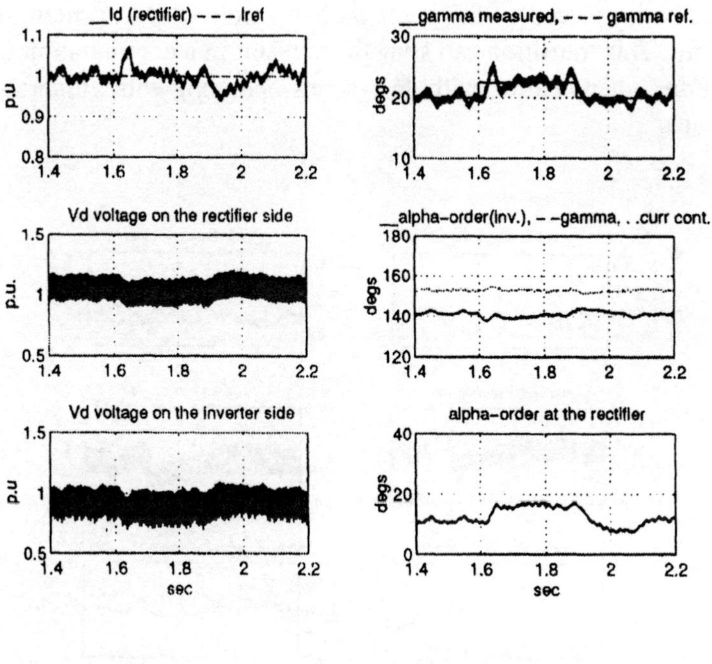

Figure 10-8: 2.5° step change in gamma reference at the inverter

10.4.2 Mode Shift

This test is a particularly interesting test to study the dynamics of the dc system when the control mode is shifted from the rectifier current controller to the inverter current controller, and vice-versa (Figure 10-9). Initially, the rectifier current controller is in control at the rectifier, whilst the gamma controller is in control at the inverter. At t = 1.1s, the ac voltage at the rectifier end is reduced by 0.05 pu which causes the rectifier to hit its alpha-min limit and lose control over the dc system current. The inverter current controller, which has been biased off by its current margin of 0.1 pu, is forced to take over current control to its reference value of 0.9 pu. This transition is usually not a difficult one for the dc system since the dc current is being reduced to 0.9 pu and the rectifier current controller is being forced to its hard limit.

At t = 1.5s, the ac voltage at the rectifier is returned to its nominal value of 1 pu, and the rectifier current controller is forced off its alpha-min limit and takes charge of the dc current. This transition is usually more difficult and dramatic since the current is being increased, and the rectifier current controller is being forced to come off its hard limit which normally requires more time. This transition can sometimes result in a commutation failure at the inverter when dealing with dc systems operating with vulnerable weak ac systems.

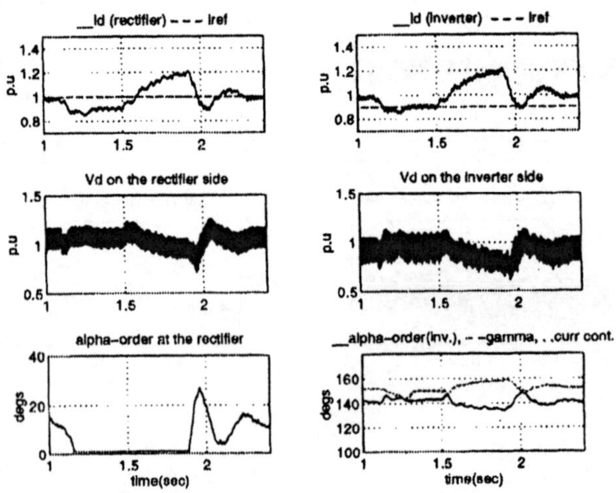

Figure 10-9: Mode shift test

10.4.3 Single-phase, 1-cycle Fault At The Inverter (Single Commutation Failure)

A commutation failure at the inverter is simulated by creating a single-phase 1-cycle fault at the inverter ac bus. The resulting commutation failure (Figure 10-10) causes the dc voltage to momentarily drop to zero and the dc current to shoot up to a value of about 1.8 pu; this current peak is limited primarily by the impedance of the dc smoothing reactor and the converter transformer impedance. This causes the dynamic portion of the Voltage Dependent Current Limit (VDCL) to operate and limit the dc current to its I_{MIN} value. The recovery process is initiated almost immediately after the fault is released, resulting in the dc current to be ramped up according to the VDCL. The alpha order at the rectifier is forced into inverter region i.e almost 145 degrees to limit the peak short circuit dc current. At the same time, the alpha order at the inverter end is forced to its alpha-min-limit of 110 degrees. During the recovery period, a transition from the inverter current controller to the rectifier current controller is also observed, as discussed previously in the case of system initialization.

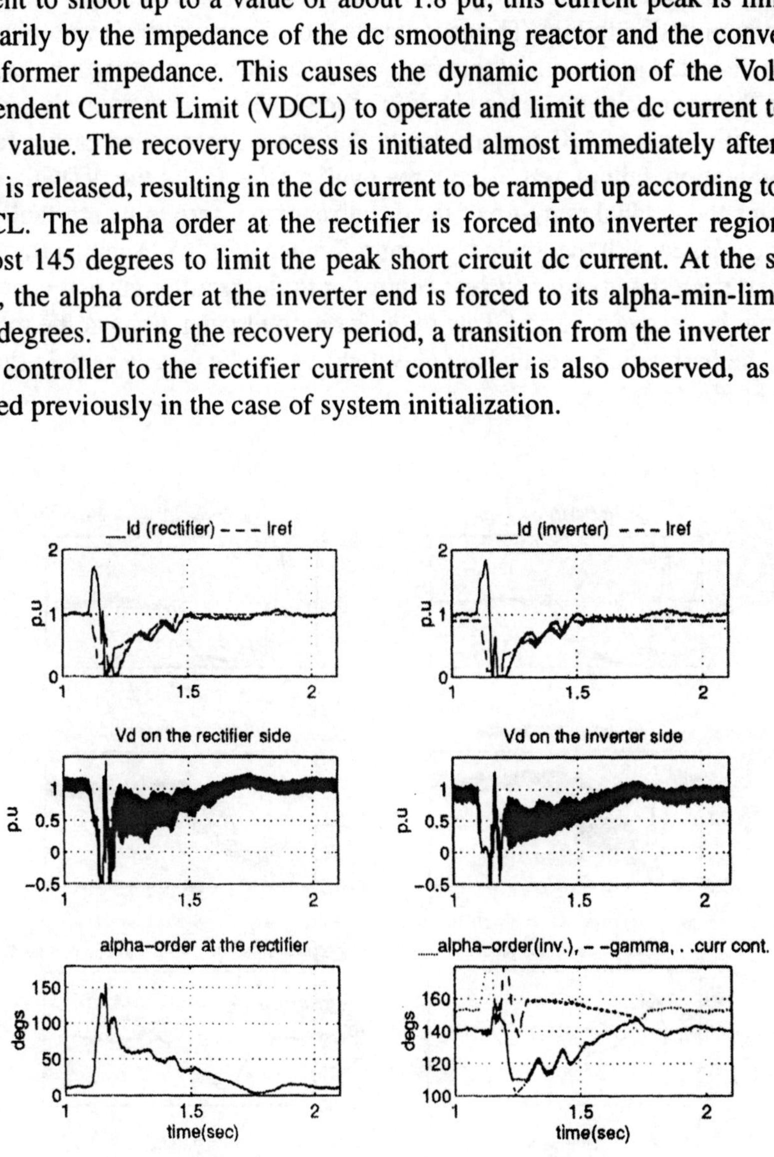

Figure 10-10: Single-Phase 1-cycle fault at the inverter (single commutation failure)

10.4.4 Single-phase 5-cycle Fault At The Inverter (Multiple Commutation Failures)

To study the dc system response for repetitive commutation failures at the inverter, a single-phase 5-cycle fault is applied at t = 1.1 s on the inverter ac bus (Figure 10-11). The initial fault causes the dc current to shoot up to 1.8 pu and collapses the dc voltage at the inverter end, as in the case of the single commutation failure. The ac voltages (not shown) at the inverter end are quite distorted as a result of this fault due to transformer saturation resulting from the over-voltages caused by loss of load. The gamma controller hits its alpha-min-limit of 110 degrees due to the reasons mentioned earlier for the commutation failure test. Upon removal of the fault, the VDCL action ensures the gradual ramp-up of the dc current from zero to its pre-fault level in accordance with the limited current reference, IOLIM. Again, a transition is seen from the inverter current controller to the rectifier current controller during the recovery period. Due to the larger operating alpha at the rectifier end, the harmonic generated are elevated and can be clearly seen in the signals of the dc voltage.

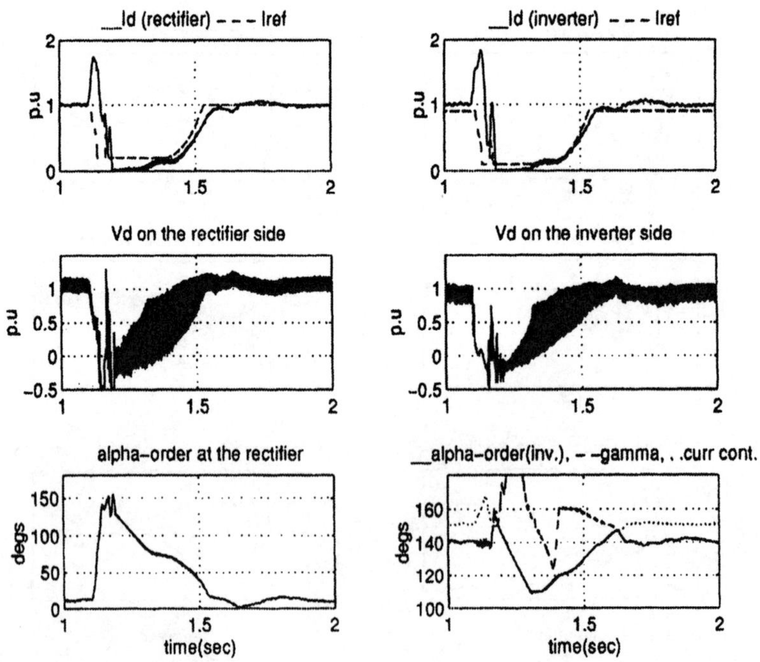

Figure 10-11: Single phase 5-cycle fault at the inverter

10.4.5 3-Phase 5-cycle Fault At The Inverter

The response of the system under this balanced three phase fault is almost identical to the response obtained during the unbalanced single-phase 5 cycle fault (Figure 10-12). The only obvious difference is seen at the inverter end dc voltage, where some oscillations are present due to the reflections on the dc transmission line. The action by the VDCL unit is able to bring the system back to steady state quickly in a controlled manner in an identical way as before.

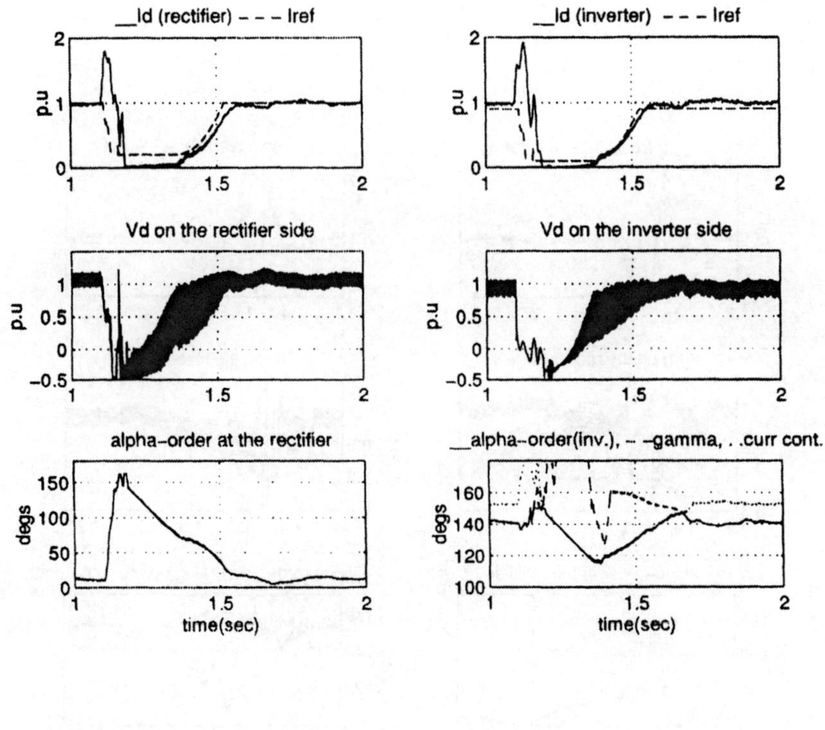

Figure 10-12: Three-phase 5-cycle fault at the inverter

10.4.6 1-phase 5-cycle Fault At The Rectifier

A single-phase 5-cycle 30% fault is applied at the rectifier ac bus (Figure 10-13). The dc current during the fault contains a second harmonic component (100 Hz, in this case) which is a characteristic of such a fault. At fault inception (at t=1.3 s), the rectifier is forced to its alpha-min limit of 9 degrees due to the voltage reduction, and the inverter current controller is forced to take over current control. Upon removal of the fault, there is a commutation failure (at t = 1.45 s). The dc current then follows a pattern similar to the single-phase one-cycle fault at the inverter. For the results shown in the CIGRE benchmark, there is a commutation failure upon recovery, and also the level of harmonic content in the dc voltage during the fault is relatively low. The difference between the results shown here and the one presented in the CIGRE benchmark [1] could be attributed to the different types of VDCL and switch models used in the two systems.

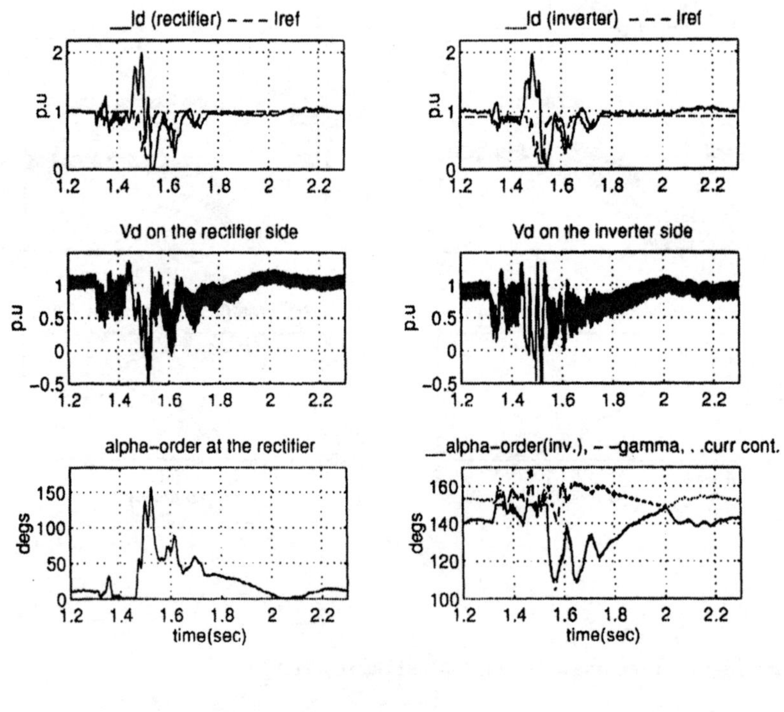

Figure 10-13: Single-phase 5 cycle fault at the rectifier

10.4.7 3-phase 5-cycle Fault At The Rectifier

The fault causes the dc voltage to collapse and that reduces the dc current to zero due to the lack of a driving source voltage. The action/behavior of the VDCL drops current order IOLIM to IMIN; however, the absence of the ac voltage at the rectifier prevents the current from recovering. Upon removal of the ac fault at the rectifier bus, the dc voltage jumps to more than 1.0 pu due to load rejection, and the current recovery follows an identical pattern (Figure 10-14) as during the initialization case. During the fault, the alpha order at the rectifier shoots up to 90 degrees followed by a gentle drop to its nominal values.

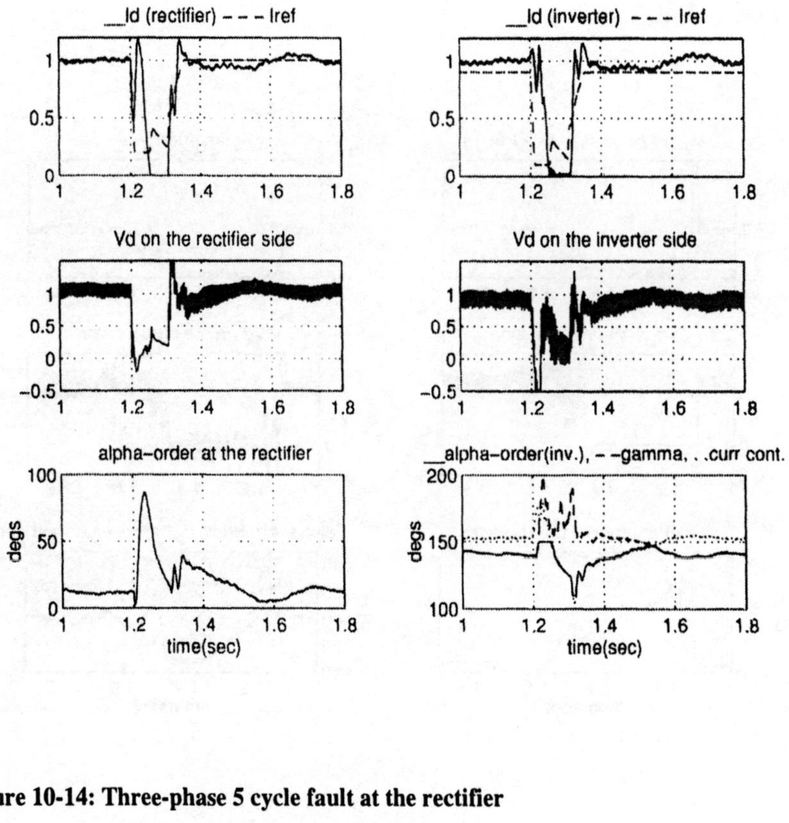

Figure 10-14: Three-phase 5 cycle fault at the rectifier

10.4.8 DC Line Fault At The Rectifier Side

A dc line fault is applied from 1.1s to 1.4s at the dc line side of the rectifier (Figure 10-15). The fault causes the dc voltage to collapse and the rectifier dc current to rise rapidly to a value greater than 2.2 pu, a characteristic of this type of fault. The peak value of the dc fault current is limited only by the size of the smoothing reactor. Since the dc fault is located at the rectifier side, the dc voltage at the rectifier is zero during the fault period. However, dc voltage reflections are observed at the inverter end. The subsequent action of the VDCL causes the current to be limited to IMIN value. From the period 1.3 s to 1.4 s, a protection signal ALPRET is applied to force the rectifier alpha to 145 degrees i.e. into inverter region to extinguish the fault current and deionize the fault arc. At the same time, the inverter is forced to its alpha-min limit of 110 degrees. Upon removal of the fault, the dc current ramps up due to the recovery action of the VDCL. The small delay observed between I_d and limited I_{ref} (IOLIM) is due to the use of a low-pass filter in the VDCL unit.

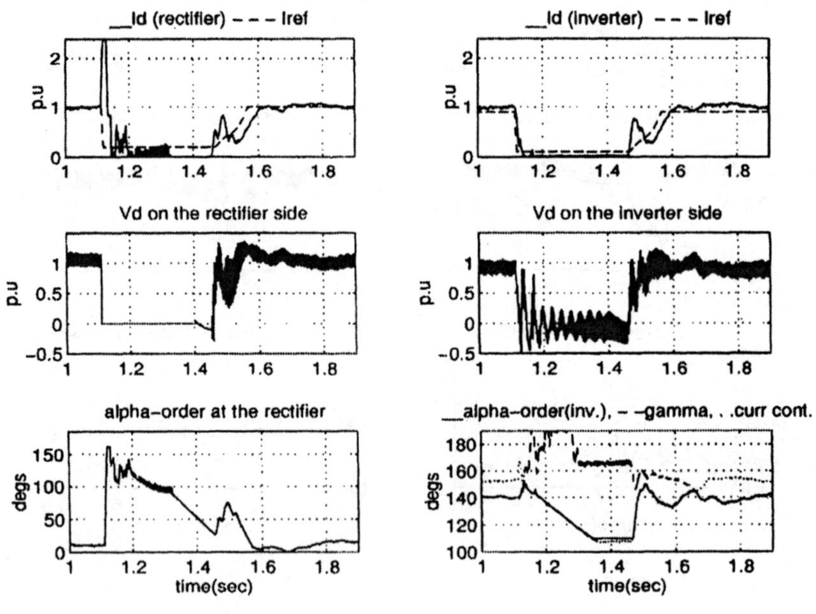

Figure 10-15: DC line fault at the rectifier side

10.4.9 DC Line Fault At The Inverter Side

A dc fault is applied from 1.1 s to 1.4 s at the dc line side of the inverter (Figure 10-16). The fault causes the dc voltage to collapse and the dc current at the rectifier to rise to a value greater than 1.8 pu This peak value is limited by the smoothing reactor and dc line impedance. Since the dc line in the inverter side is shorted out by the dc line fault, the dc voltage at the inverter end is zero during the fault period. However, voltage reflections are observed in the dc line voltage at the rectifier end. The subsequent action of the VDCL causes the dc current order to be limited to minimum current (IMIN). The alpha order at the rectifier reacts to the short circuit and increases from its pre-fault (nominal) value of about 17 degrees to a maximum value of about 145 degrees in order to reduce the fault current. From 1.3 s to 1.4 s a protection signal, alpha-retard (ALPRET), is applied to deionize the fault arc. The alpha order at the inverter is forced to its alpha-min-in-inverter mode setting of 110 degrees. Upon removal of the fault, the dc current ramps up due to the action of the sloping portion of the VDCL. The small delay observed between I_d and limited current order, IOLIM, is due to the use of a low-pass filter in the VDCL unit.

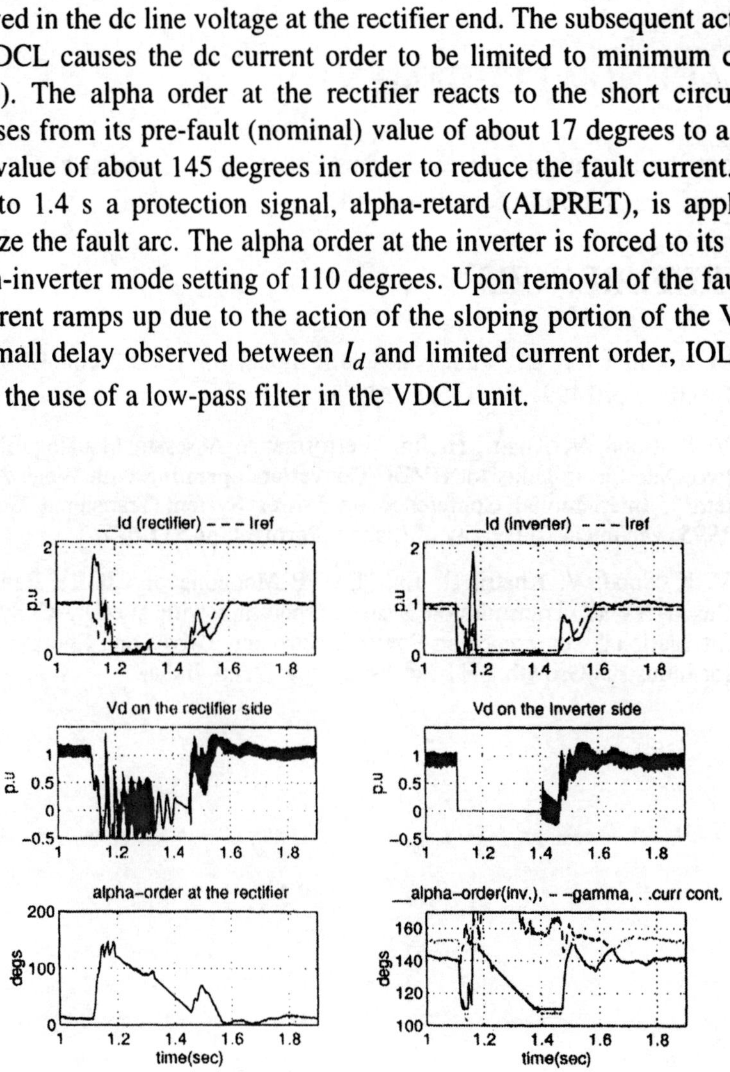

Figure 10-16: DC line fault at the inverter side

10.5 CLOSING REMARKS

Results from an EMTP-based simulation of a modified 6-pulse version of the CIGRE HVDC benchmark have been presented in this chapter. A model of the control system has been presented which contains typical features found in most commercial HVDC control systems. The kinds of results expected and comments that describe the features to look for have been presented.

10.6 ACKNOWLEDGEMENT

The contributions of my former student Mr. V. Khatri whilst at Concordia University are gratefully acknowledged.

10.7 REFERENCES

[1]. M.Szechtman et al, "First benchmark model for HVDC control studies", Electra, April 1991, No.135, pp 55-73.

[2]. V. K. Sood, V. Khatri, H. Jin, "Performance Assessment using EMTP of two Gate Firing Units for HVDC Converters operating with Weak AC Systems", International Conference on Power System Transients, Sept. 3-7 1995, Technical University of Lisbon, Portugal, pp 517-522.

[3]. V. K. Sood, V. Khatri, H. Jin, "EMTP Modeling of CIGRE Benchmark Based HVDC Transmission System Operating with Weak AC Systems", International Conference on Power Electronics, Drives and Energy Systems for Industrial Growth, 8-11 Jan 1996, New Delhi, India.

Chapter 11

Advanced Controllers

11.1 INTRODUCTION

In recent years, development of modern control techniques has speeded up and the understanding of these new controls has improved. Utility engineers are now just beginning to consider the possibilities offered by such modern control techniques to assist in the operation and stabilization of power systems [1]. However, the utility industry is conservative and new control techniques will only be adopted with extreme caution and only when conventional controllers are found unsuitable for the task at hand.

The whole subject of intelligent controllers was given a boost by the seminal paper by K. Narendra and K. Parthasarathy in 1990 [3]. This was followed by numerous publications dealing with controllers based on adaptive gain-scheduling techniques [6], neural networks [7,8], fuzzy logic [5,8,9-15] and other optimal techniques.

Since HVDC systems are fast acting, and can be controlled within tens of milli-seconds, it is feasible that advanced controllers will be used with HVDC systems before other power system applications. In the past ten years, much work has been published on the use of fuzzy logic based controllers [9-15].

In HVDC systems, the use of a Voltage Dependent Current Limit (VDCL) function was a crude yet effective method to adapt the current reference to the prevailing system conditions, especially during dynamic system recovery conditions. This chapter deals with a more elegant VDCL based on neural networks and fuzzy logic.

11.2 APPLICATION OF AN ADVANCED VDCL UNIT

11.2.1 Introduction

It is well known that an HVDC converter feeding into a weak ac system is prone to commutation failures. Under such conditions, an adaptive current (or power) reference [1] can be useful to optimize the system recovery following a fault and alleviate the possibility of subsequent commutation failures.

In HVDC transmission systems, a VDCL unit has been traditionally used to generate an adaptive current reference for the converter controller. This current reference can be adapted based on either (a) the dc (transmission line) voltage or (b) the (rectified) ac voltage at the filter bus of the converter. The choice as to which of these two voltages is used is a function of the desired system stability and performance following any perturbations or faults. The traditional VDCL unit [12] is a Multiple-Input Single-Output (MISO) type with a non-linear voltage-current (input-output) characteristic. The VDCL unit is composed of two sub-units having static and dynamic characteristics (Figure 11-1).

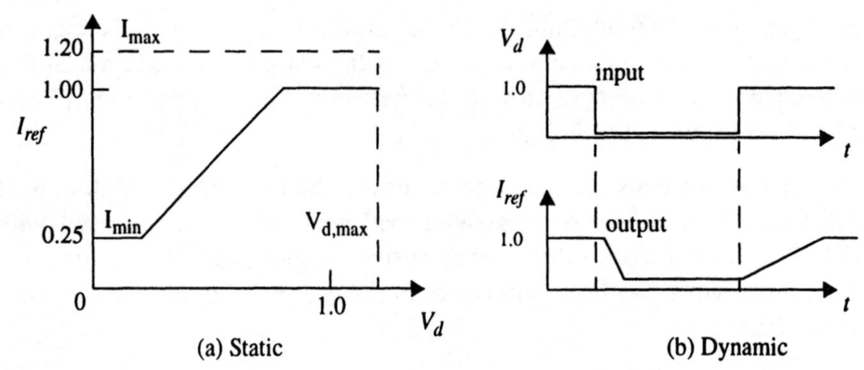

Figure 11-1: Characteristics of the VDCL

In the proposed Neuro-Fuzzy (NF) VDCL unit, multiple inputs (i.e. dc voltage, dc current references) are used to produce an output adaptive current-reference. To arrive at the control action, a novel fuzzy centroid inference algorithm is used directly on the output of a Radial Basis Function (RBF) Neural Network (NN). This new VDCL unit is tested with a simplified model of a two terminal HVDC system which is suitable for demonstrating the preliminary dynamic analysis. Some simulation results of the NF VDCL unit are presented and analyzed.

11.2.2 Fuzzy Inference

In a Fuzzy system (Figure 11-2), a crisp input X is fuzzified using a Fuzzifier. The Fuzzifier output is fed to the Fuzzy Inference Engine which operates via a Fuzzy rule-based system. The output Y from the Fuzzy Inference Engine is defuzzyfied and then converted into a crisp output Z amenable for control action. Most of these functions (enclosed by the dotted line in Figure 11-2) are performed by a RBF NN.

If the sum of the outputs of the Gaussian hidden layer is unity, then their outputs can be combined linearly with the weights of the output layer. In the present application, the condition of unity on the sum is not imposed and hence normalization has to be done on the output of the RBFNN.

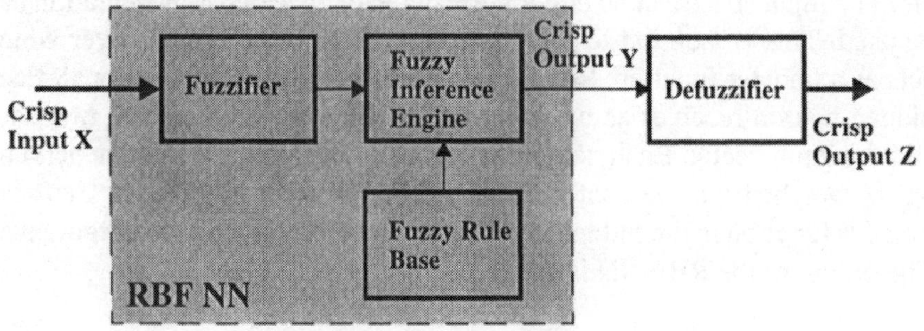

Figure 11-2: Fuzzy system interpretation with RBFNN

A new inference mechanism, based on a Fuzzy Centroid formula, is used on the output of the RBF NN to arrive at a meaningful control action. This output is actually considered as the contribution of the membership function formed by a set of fuzzy linguistic statements on the Universe of Discourse of the output space, $z = \{z_1, z_2 .., z_n\}$. The defuzzified output of the ith pattern of the system is then given by:

$$Z = \Sigma(\mu_i z_i)/\Sigma(\mu_i) \qquad (11\text{-}1)$$

where:

Z = Output of the defuzzifier,

z_i = Point in the output space,

μ_i = Output of the RBF NN.

The above method of utilizing the RBF NN as a fuzzy system is possible because of (a) the functional equivalence [2] between them, and (b) the ability of the scaled Gaussian membership function to universally approximate any continuous function.

In the present study, the dc voltage V_d and seven current order characteristics $I_{o1},...I_{o7}$, each defined over 12 points (Figure 11-4) are the inputs fed to the RBF NN. The RBF NN then produces a single adaptive current reference, I_{ref} as an output after defuzzyfication.

11.2.3 Structure of RBF NN

A typical RBF NN structure (Figure 11-3) has input, hidden and output layers. The input space can be either normalized or an actual representation can be used. This is then fed to the associative cells of the hidden layer which acts as a transfer function. Representing bias in these cells is optional. Each hidden neuron receives as net input the distance between its weight vector and the input vector. Each neuron in the RBF NN outputs a value depending on its weight from the center of the RBF. The RBF NN uses a Gaussian transfer function in the hidden layer and a linear function in the output layer. The output of the RBF NN is given as:

$$O_k = exp\left(\frac{-[x - c_k]^T[x - c_k]}{2\sigma^2{}_k}\right) \qquad (11\text{-}2)$$

where:

k = 1,2,...,N (where N=84, is the number of hidden nodes),

O_k = output of the kth node of the hidden layer,

x = input pattern vector,

$c_k =$ center of the RBF of kth node of the hidden layer,

$\sigma_k =$ spread of the kth RBF.

The output of the jth node is given by:

$$y_j = \mathbf{W}_j^T O_j \quad (11\text{-}3)$$

where:

$j =$ 1,2,....,M (where M=84, is the number of output nodes)

$y_j =$ output of the jth node,

$\mathbf{W}_j =$ weight vector for node j,

$O_j =$ vector output from the jth hidden layer (can be augmented with bias vector).

Choosing the spread of the RBF, depends on the pattern to be classified. Many algorithms are available to find the optimal values of centers and spread of the RBF [3,4,5]. Generally, the spread should be larger than the minimum distance and smaller than the maximum distance between the input vector and the center of the RBF spread to get better generalization. The linear coefficients of the output layer are the adjustable weights **W**, and since the output is linearly dependent on the input set, the solution is obtained by solving a linear optimization problem. In this paper, the center of the RBF and the weights are found using the orthogonal least squares (OLS) algorithm [3].

Defuzzified output is obtained by substituting y_j for μ_j in equation 1.

The advantages of using a Gaussian RBF are:

- RBFs are functionally equivalent to Fuzzy systems [2],
- Since the hidden and output layer parameters can be independently evaluated, training is faster [6],
- A single hidden layer is sufficient to approximate the given function [7],
- The RBF parameters have a close relationship with the sampling theorem and hence stable control of the system is possible [8], and

- The RBF NN architecture is easy to implement using VLSI techniques [9].

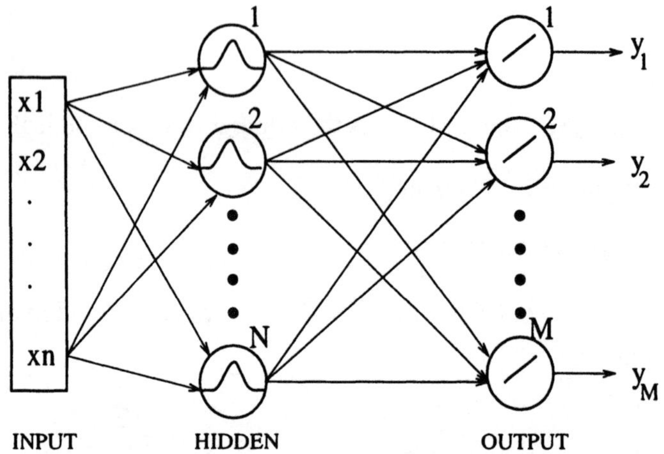

Figure 11-3: Typical Gaussian Radial Basis Function Neural Network

As an example, the RBF NN capability to reproduce a sine-wave (amplitude, U), given only 5 input points (0,1,0,-1,0), is compared (Figure 11-4) with that of three other frequently used algorithms:

- Linear interpolation (look-up table),
- Cubic spline, and
- Two-layer feed-forward NN using the Levenberg optimization.

The input sampling rate is 200 Hz and the recall performance is verified at a 50 kHz sampling rate. The error plot of the capability of the various methods in reproducing the sine-wave shows that the RBF NN has the lowest error when compared with the three other methods. This is because of the localization effect of the Gaussian RBF due to which the NN will have a maximum output when the input pattern is close to the center of the RBF.

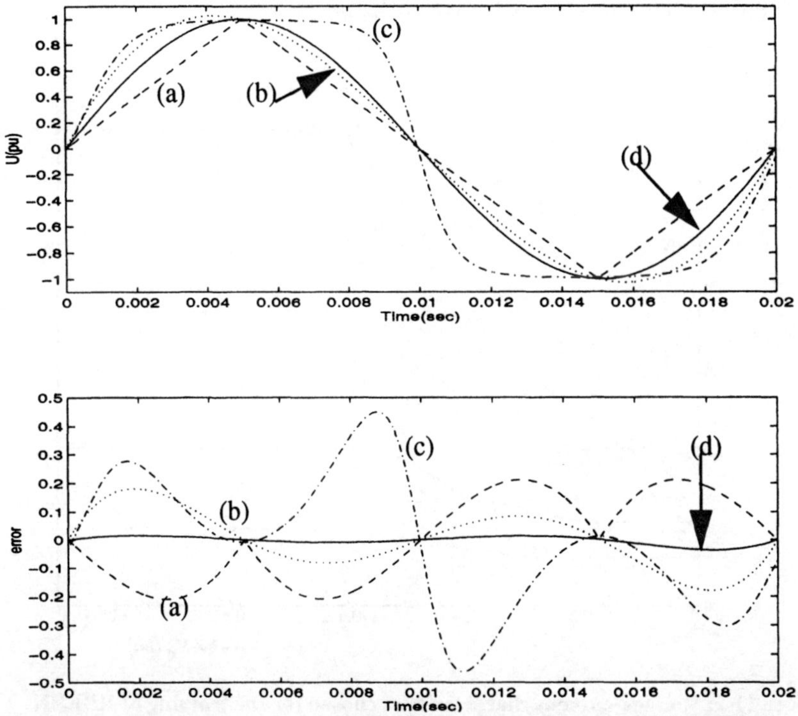

(a) Linear interpolation (b) Cubic spline (c) Levenberg NN (d) RBFNN

Figure 11-4: Comparative analysis of RBF capability to reproduce a sine wave

11.2.4 Methodology

The per-unit voltage-current characteristics to be fed to the RBF NN are shown in Figure 11-5. It consists of multiple current reference characteristics instead of a single current reference (Figure 11-1) to improve the system performance especially at low dc voltages due to faults.

The input pattern is classified into 8 variables composed of 7 current orders (I_{o1},...,I_{o7}) and the dc voltage, V_d. As described in the previous section, the interpolation capability of the RBF NN in reconstructing the unknown function is effective even with only a few input points defined over the input space. Hence, each current order and the dc voltage are defined over only 12 points.

The output is divided into 12 patterns in 7 variables corresponding to each characteristic. One of the sample input/output patterns to the RBF NN used for the first characteristic is shown in Table 11-1.

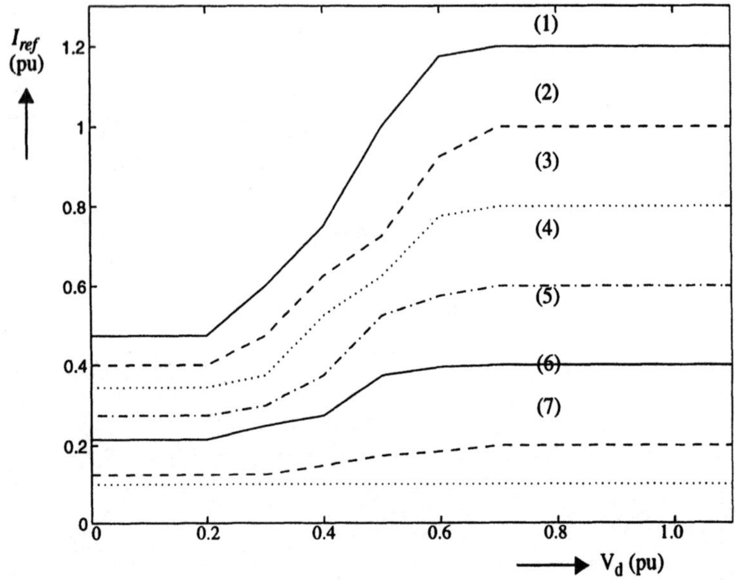

Figure 11-5: Voltage-current characteristics chosen for the training of RBFNN

Table 11-1: Sample input-output pattern classification of the RBF NN

	INPUT VARIABLES								OUTPUT
#	I_{o1}	I_{o2}	I_{o3}	I_{o4}	I_{o5}	I_{o6}	I_{o7}	V_d	I_{ref}
1	1	0	0	0	0	0	0	0	0.475
2	1	0	0	0	0	0	0	0.1	0.475
3	1	0	0	0	0	0	0	0.2	0.475
4	1	0	0	0	0	0	0	0.3	0.600
5	1	0	0	0	0	0	0	0.4	0.750
6	1	0	0	0	0	0	0	0.5	1.000
7	1	0	0	0	0	0	0	0.6	1.175
8	1	0	0	0	0	0	0	0.7	1.200
9	1	0	0	0	0	0	0	0.8	1.200
10	1	0	0	0	0	0	0	0.9	1.200
11	1	0	0	0	0	0	0	1.0	1.200
12	1	0	0	0	0	0	0	1.1	1.200

The RBF NN is trained off-line using the OLS algorithm [3] and used on-line to perform the control action.

11.2.5 HVDC System Considered For The Study

11.2.5.1 HVDC system

The HVDC system used in this study, derived from the CIGRE bench-mark model [10], is a quasi-steady state model with the simplifications that (a) the rectifier is a variable dc voltage source with a 12-pulse ripple superimposed on it, (b) the inverter is an ideal dc voltage source, and (c) the dc system is an equivalent transfer function.

Since the converters are assumed ideal, no commutation transients are represented. The derived simple transfer function model of the plant was simulated using the Matlab SIMULINK software package to permit conceptual insights into the controller behavior. In later investigations it is intended to replace this dc model with a more realistic HVDC system model.

11.2.5.2 Control system representation

In the proposed Neuro-Fuzzy (NF) VDCL unit (Figure 11-6), multiple inputs (i.e. dc voltage, 7 current orders) are fed to the RBF NN via a switch which is used to select either the manual or adaptive input. In the manual mode, a desired current reference characteristic can be set for the RBF NN whereas in the adaptive mode, the RBF NN will produce simultaneously 7 outputs and adapts only one output depending on the dc voltage.

The output of the RBF NN is fed to the centroid defuzzifier to get the single adapted current reference I_{ref} current reference which is then fed to a traditional PI controller.

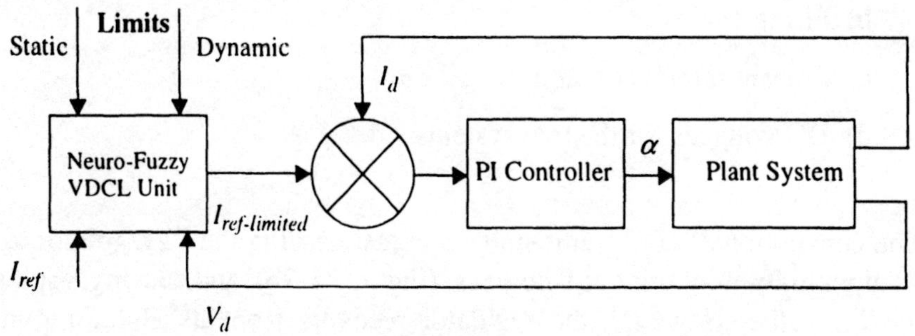

Figure 11-6: Control system schematic with Neuro-Fuzzy VDCL unit

11.2.6 Results And Discussions

The performance of the proposed Neuro-Fuzzy VDCL unit is evaluated by simulating the following four case studies:

1. Starting-up of the dc system,
2. Reduction of dc voltage,
3. Recovery from fault, and
4. Current reference tracking.

The results are compared for the two systems having either (a) a conventional VDCL unit, or (b) a Neuro-Fuzzy (NF) VDCL unit.

As described earlier, the NF VDCL unit is equipped with different voltage-current characteristics and the adaptive current reference I_{ref} given out from this unit depends on the dc voltage and the current order setting at the local terminal. The conventional VDCL unit has a single voltage-current characteristic generated by a ramp function which is simulated as a "look-up" table. For both systems, the PI controller used has identical controller gain parameters.

11.2.6.1 Case 1 - Starting-up Of DC System

The selection of appropriate control system parameters is very important to the start-up performance of the dc system.The dc-link is started from zero initial conditions and the dynamic response of the system is shown in Figures 11-7a,b,c & d.

The following signals are shown in the figure:

a) DC currents,
b) Firing angles,
c) Current references, and
d) DC voltages for the two systems.

The conventional VDCL unit start-up is fast causing the PI regulator to hit its alpha-minimum limit of 9 degrees. (Figure 11-7b), and causing a spike in the I_{ref} value. However, the regulator recovers from its alpha-minimum limit at 0.01 s and quickly reaches its final value of alpha = 16 degrees at

0.125 s. The dc current recovery is smooth all the way, attaining 90% of its value in 0.2 s.

The NF VDCL unit start-up is slightly better controlled and does not hit its alpha-minimum limit at all. It attains 90% of its final dc current value at practically the same time as the conventional VDCL unit.

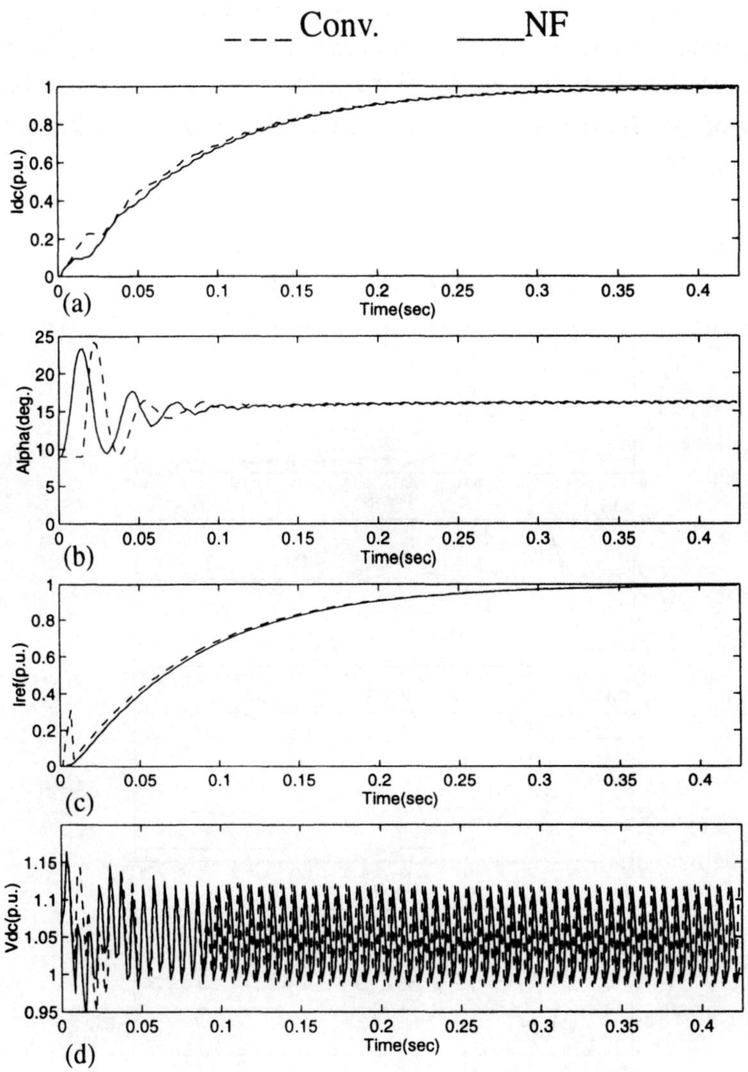

Figure 11-7: Performance during starting-up of dc system

11.2.6.2 Case 2 - Reduction Of DC Voltage

To check the dynamic performance of the HVDC system, the dc voltage at the inverter terminal is transiently reduced to zero to simulate the effect of a 3-phase fault at ac bus of the inverter. The results are depicted in Figures 11-8 a, b, c & d corresponding to dc currents, firing angles, current references and dc terminal voltages at the rectifier-end, respectively.

Both conventional and NF VDCL units are able to reduce their current references within 1-2 cycles to their limited values; 0.4 pu in the case of the conventional VDCL unit and 0.1 pu in the NF VDCL case. Moreover, the NF unit is slightly faster (see I_{ref} signals). A characteristic oscillation frequency of 100 Hz (second harmonic on the CIGRE benchmark system) is also observed.

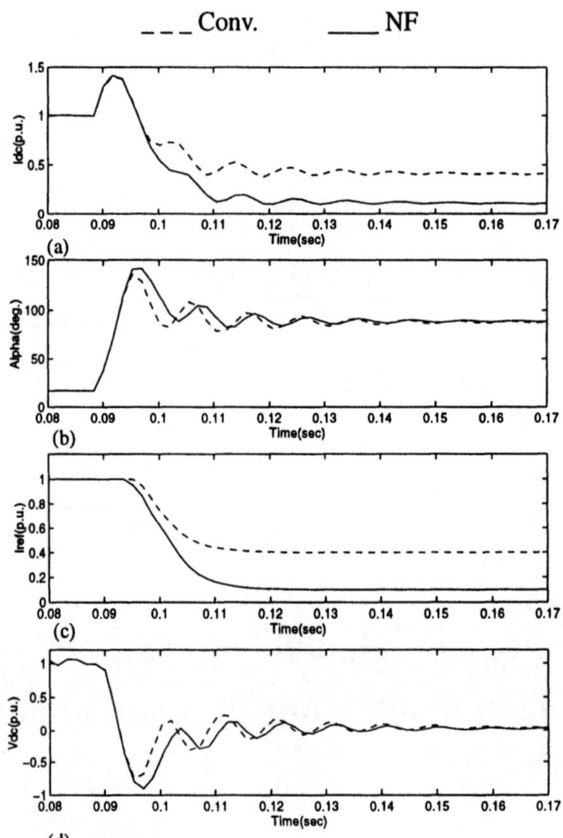

Figure 11-8: System response during fault

11.2.6.3 Case 3 - Recovery From Fault

In this case, the system recovery from a fault is considered assuming that the system dc voltage has re-established to 0.5 pu; this value is chosen since it falls within the sloped region of the voltage-current characteristics (Figure 11-5) permitting examination of the VDCL sensitivity. The responses of the two systems are presented in Figures 11-9a, b, c and d.

Here, the conventional VDCL system is sensitive to the sloped region and, as a result, it produces an oscillatory response (Figures 11-9 a, b, c and d) compared to the NF case which provides a more damped response than the conventional case. Since the conventional VDCL unit I_{ref} oscillates to within 0.8 pu, there is a possible danger from a commutation failure for this system. The NF VDCL unit exhibits no such oscillations.

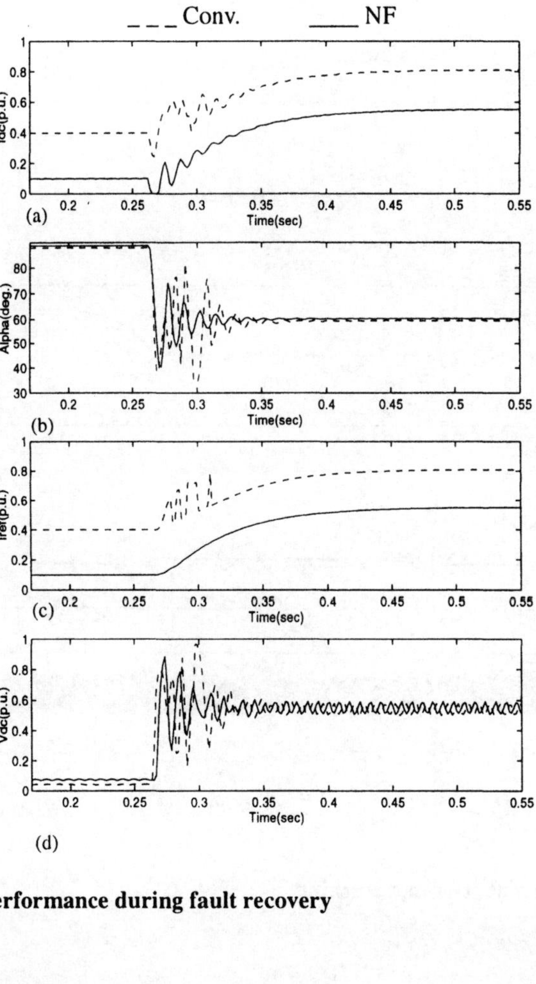

Figure 11-9: Performance during fault recovery

11.2.6.4 Case 4 - Current Reference Tracking

HVDC systems are well-known for their fast controllability to carry the desired dc power, or to modulate the dc power to improve the stability of an attached ac system. One measure of fast controllability is usually verified by considering the current reference tracking performance of the dc controller. This test is carried out by reducing the current reference manually by 10% or so in a practical system.

In this particular instance, a 20% step change in the current reference (Figure 11-10b) to 0.8 pu is initiated for a period of 25 cycles starting at 0.8 s. The results are compared for a case with no VDCL (i.e. completely unlimited linear case) and the NF VDCL in place. The resulting dc current, the current orders, the firing angles and the terminal dc voltages are shown in Figures 11-10 a, b, c and d respectively.

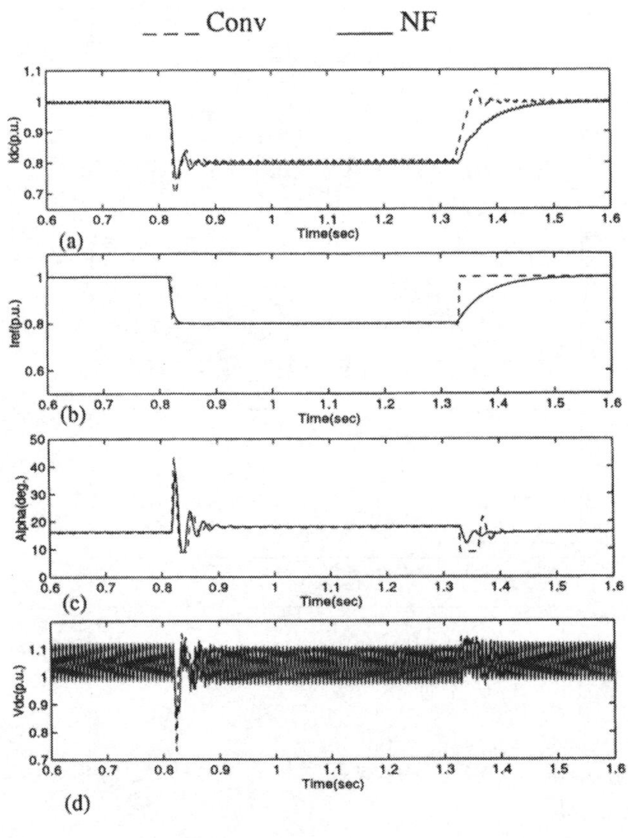

Figure 11-10: Current reference tracking

It can be seen that the NF system has a damped tracking ability with no oscillations or overshoot. Certainly, the I_{ref} value could be set to provide a similar performance but it would require considerable optimization of the controller parameters.

11.3 CONCLUSIONS

A new method of combining an RBF NN with a fuzzy inference mechanism to produce an adaptive current reference for the VDCL unit of an HVDC controller is proposed. Preliminary results from the proposed Neuro-Fuzzy VDCL unit show that it can enhance the performance of an HVDC system under dynamic operating conditions.

Further work is needed to test the proposed Neuro-Fuzzy VDCL unit with the following:

- More detailed and realistic representation of the HVDC system,
- NN controller [11] instead of a conventional PI controller, and
- MTDC system where the adaptive VDCL characteristics can have a significant role to play.

11.4 ACKNOWLEDGEMENT

The contributions of my former associate Mr. K.Narendra are gratefully acknowledged.

11.5 REFERENCES

[1]. S.Lefebvre, M.Saad, and A.R. Hurteau, "Adaptive control for HVDC power transmission systems", IEEE Trans. Power Apparatus Systems, Vol. PAS-104, No.9, Sept. 1985, pp 2329-2335

[2]. Y.Hsu and C.Cheng, "Design of fuzzy power system stabilizer for multi-machine power system", Proc. IEE, May 1990, 137, pp 233-238

[3]. K. S. Narendra and K. Parthasarathy, "Identification and Control of Dynamical Systems using Neural Networks", IEEE Trans. Neural Networks, Vol. 1, no. 1, pp. 4-27, Mar. 1990.

[4]. R. Jayakrishna, "Application of Knowledge Based Controls for Enhancing the Performance of an MTDC-AC System", Thesis, Indian Institute of Science, India, Dec. 1993.

[5]. P.K.Dash, A.Routray and S.Rahman, "An adaptive fuzzy logic controller for ac-dc power systems," Paper 0-7803-1217-1, IEEE 1993

[6]. J.Reeve and M.Sultan, "Gain scheduling adaptive control strategies for HVDC systems to accommodate large disturbances", IEEE Trans. Power Systems, Vol.9, No.1, Feb. 1994, pp 366-372

[7]. V. K. Sood, N. Kandil, R. V. Patel & K. Khorasani, "Comparative Evaluation of Neural-Network-Based and PI Current Controllers for HVDC Transmission", IEEE Trans. Power Electronics, Vol. 9, no.3, pp. 288-296, May 1994.

[8]. B. K. Bose, "Expert System, Fuzzy Logic, and Neural Network Applications in Power Electronics and Motion Control", Proc. IEEE, Vol. 82, no. 8, pp. 1303-23, Aug. 1994.

[9]. P. K. Dash, A. C. Liew, A. Routray, "High Performance Controllers for HVDC transmission links", IEE Proc. Gen. Trans. and Distrn. Vol. 141, No.5, pp. 422-428, Sept. 1994.

[10]. P.K.Dash, A.C. Liew and A.Routray, "High performance controllers for HVDC transmission links", IEE Proc. Gener. Transm. Distr., Vol.141, No.5, September 1994.

[11]. K.G. Narendra, V.K.Sood, R.V.Patel and K.Khorasani, "Neuro-Fuzzy VDCL unit to enhance performance of HVDC system", Canadian Conference on Electrical and Computer Engineering, Montréal, Sept., 1995.

[12]. P.K.Dash, A.Routray and S.K.Panda, "Gain scheduling adaptive control strategies for HVDC systems using fuzzy logic", Paper No. 0-7803-2795-0 IEEE, pp 134-139.

[13]. P.K.Dash, A.Routray, S.K.Panda and A.C.Liew, "Fuzzy tuning of dc link controllers", IEEE Catalogue No. 95TH8130, Paper No. 0-7803-2981-3, 1995, pp 370-375.

[14]. P.K.Dash, A.Routray, S.K.Panda, "A fuzzy self-tuning PI controller for HVDC links", IEEE Trans. on Power Electronics, Vol. 11, No.5, September 1996, pp 669-679.

[15]. A.Daneshpooy, A.M.Gole, D.G.Chapman and J.B. Davies, "Fuzzy logic control for HVDC transmission", IEEE Trans. on Power Delivery, Vol. 12, No.4, October 1997, pp 1690-1697.

Chapter 12

Measurement/Monitoring Aspects

12.1 INTRODUCTION

Some aspects of the monitoring that is required to provide measured data to the controllers will be discussed in this chapter. The data is from various manufacturers (i.e. Siemens and ABB) for their latest equipment dealing with:

- Current / Voltage transducers: ac and dc,
- Firing Angle Measurement, and
- Fault Diagnosis.

12.2 MONITORING OF SIGNALS

The monitoring of the following signals is necessary for the controllers (Figure 12-1) to perform their functions and assist in the protection of the converter equipment:

- V_d, I_d - dc voltage and dc current respectively,
- I_{ac}, I'_{ac} - ac current on the line- and converter-side of the converter transformer,
- V_{ac} - ac voltage at the ac filter bus, and
- Valve conduction status.

The three phase ac voltage at the commutation (filter) bus can be measured with the help of an auxiliary transformer. Due to the presence of ac filters at this bus, the voltage waveform is devoid of commutation notches, but may contain some level of harmonic content, particularly if the ac system is considered weak.

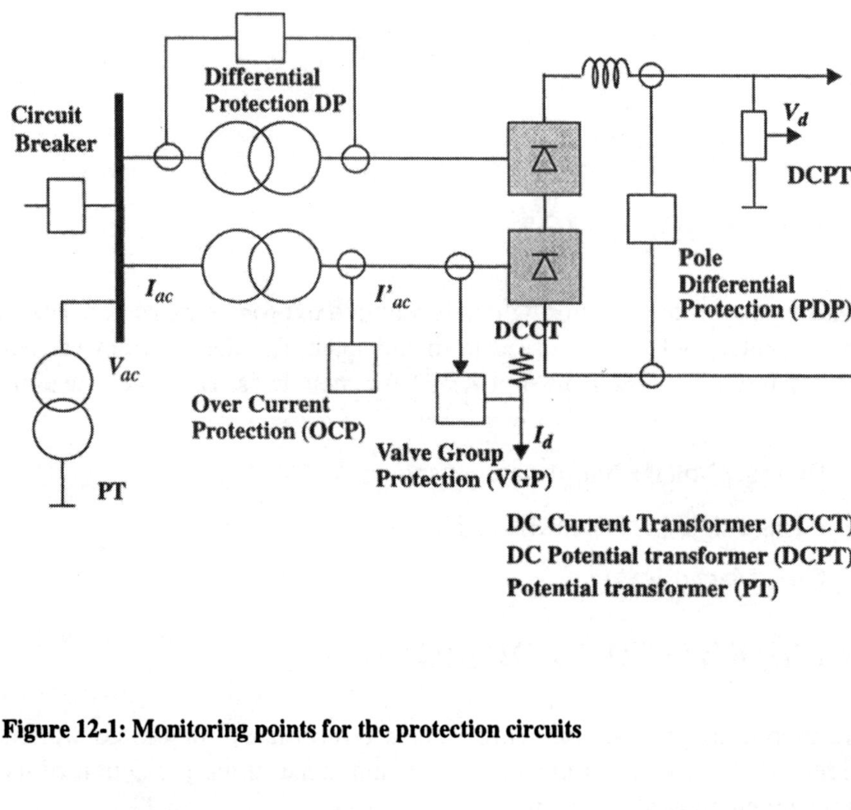

Figure 12-1: Monitoring points for the protection circuits

The dc voltage is often measured by means of either bushing-type capacitive or resistive dividers.

The dc current used to be measured with transductor type dc current transformers. However, these are now superseded and replaced by a precision shunt at high potential, and the use of an optical fibre link between ground and high potential. These new transducers can handle both ac and dc current measurements.

These new optical current transducers utilize a precision shunt at high potential. A small optical fibre link between ground and high potential is utilized resulting in a lower probability of a flashover. The optical power link transfers power to high potential for use in the electronics equipment. An optical Data Link transfers data to ground potential. This transducer results in high reliability, compact design and efficient measurement with accuracy of 0.5% within the frequency range from dc to 7 kHz (see also chapter 15).

The measurement of valve conduction status is necessary for both protection and extinction angle control at an inverter. The use of fibre optics is made to transmit the valve (and individual thyristor) conduction status to ground potential. The timing of the valve extinction is obtained by comparison with the appropriate ac voltage zero crossing.

12.3 PROTECTION AGAINST OVER-CURRENTS

Faults and disturbances can be caused by malfunctioning equipment or insulation failures due to lightening and pollution. First, these faults need to be detected with the help of monitored signals. Second, the equipment must be protected by control or switching actions. Since dc controls can react within 1 cycle, control action is used to protect equipment against over currents and over voltage stresses, and minimize loss of transmission. In a converter station, the valves are the most critical (and most expensive) equipment that need to be protected rapidly due to their limited thermal inertia.

The basic types of faults that the station can experience are:

- **Current Extinction (CE)**

  CE can occur if the valve current drops below the holding current of the thyristor. This can happen at low current operation accompanied by a transient leading to current extinction. Due to the phenomena of current chopping of an inductive current, severe over-voltages may result. The size of the smoothing reactor and the rectifier I_{min} setting helps to minimize the occurrence of CE.

- **Commutation Failure (CF) or misfire**

  In line-commutated converters, the successful commutation of a valve requires that the extinction angle γ-nominal be maintained more than the minimum value of the extinction angle γ-min. Note that γ-nominal = 180 - α - μ. The overlap angle μ is a function of the commutation voltage and the dc current. Hence, a decrease in commutation voltage or an increase in dc current can cause an increase in μ, resulting in a decrease in γ. If $\gamma < \gamma$min, a CF may result. In this case, the out-going valve will continue to conduct current and when the in-coming valve is fired in sequence, a short circuit of the bridge will occur.

A missing firing pulse can also lead to a misfire (at a rectifier) or a CF (at an inverter). The effects of a single misfire are similar to those of a single CF. Usually a single CF is self-clearing, and no special control (other than controller response) actions are necessary. However, a multiple CF can lead to the injection of ac voltages into the dc system. Control action may be necessary in this case.

The detection of a CF is based on the differential comparison of dc current and the ac currents on the valve side of the converter transformer. During a CF, the two valves in an arm of the bridge are conducting. Therefore, the ac current goes to zero while the dc current continues to flow.

The protection features employed to counteract the impact of a CF are indicated in Table 12-1.

Table 12-1: Protection against over-currents

Fault type	Occurrence	Fault current level	Protection method
Internal faults	Infrequent	10 pu	Valve is rated to withstand this surge
DC line faults	Frequent	2-3 pu	- forced retard of firing angle - dynamic VDCL deployment - trip ac breaker CB after third attempt
Commutation failures (single or multiple)	Very frequent	1.5-2.5 pu	**Single CF:** - self-clearing, Multiple CF: - Beta angle advanced in stages - static VDCL deployment

- **Short Circuits - internal or dc line**

An internal bridge fault is rare as the valve hall is completely enclosed and its air-conditioned. However, a bushing can fail, or valve cooling

water may leak resulting in a short circuit. The ac breaker may have to be tripped to protect against bridge faults.
The protection features employed to counteract the impact of short circuits are indicated in Table 12-1.

The fast-acting HVDC controls (which operate within 1 cycle) are used to regulate the dc current for protection of the valves against ac and dc faults.

The basic protection (Figure 12-2) is provided by the VGP differential protection which compares the rectified ac current on the valve side of the converter transformer with the dc current measured on the line side of the smoothing reactor. This method is applied because of the selectivity possible due to high impedances in the smoothing reactor and converter transformer.

The OCP is used as a back-up protection in case of malfunction in the VGP. The level of over-current setting is set higher than the differential protection.

The Pole Differential Protection (PDP) is used to detect ground faults, including faults in the neutral bus.

12.4 PROTECTION AGAINST OVER-VOLTAGES

The typical arrangement of metal-oxide surge arrestors for protecting equipment in a converter pole is shown in the Figure 12-2. In general, overvoltages entering from the ac bus are limited by the ac bus arresters; similarly, overvoltages entering the converter from the dc line are limited by the dc arrestor. The ac and dc filters have their respective arrestors also. Critical components such as the valves have their own arrestors placed close to these components. The protective firing of a valve is used as a backup protection for overvoltages in the forward direction. Owing to their varied duty, these arrestors are rated accordingly for the location used. For instance, the converter arrestor for the upper bridge are subjected to a higher energy dissipation than for the lower bridge.

Since the evaluation of insulation co-ordination is quite complex, detailed studies are often required with dc simulators to design an appropriate insulation co-ordination strategy.

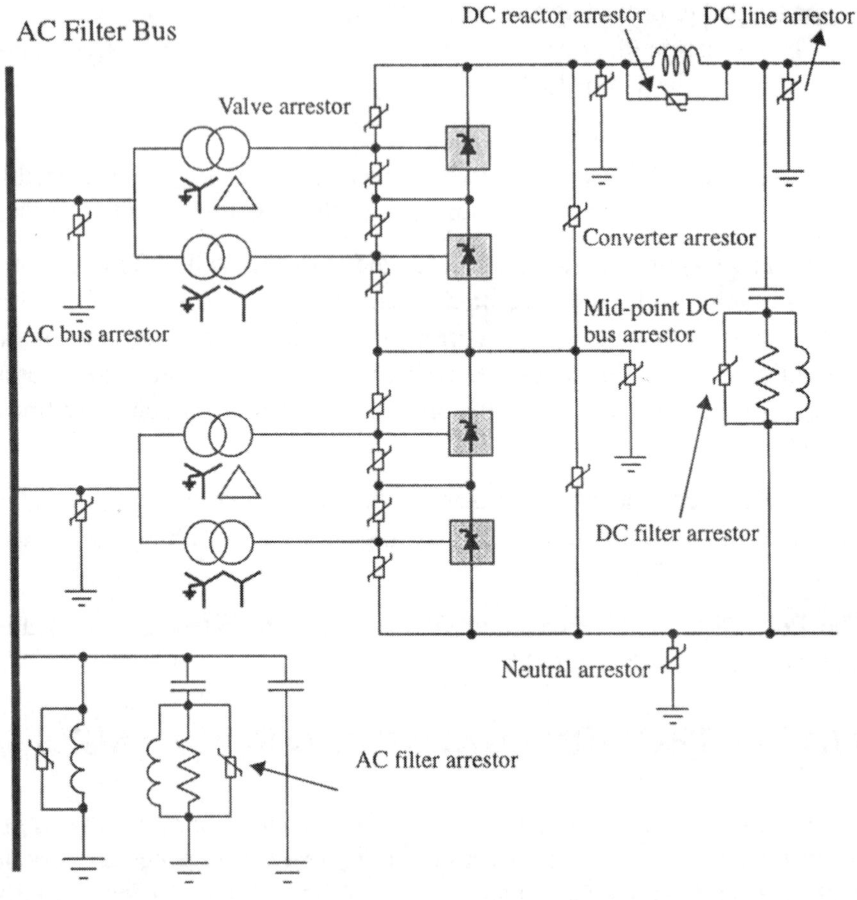

Figure 12-2: Typical arrangement of surge arrestors for a converter pole [2]

12.5 ACKNOWLEDGEMENT

The use of internal documents from ABB and Siemens is acknowledged.

12.6 REFERENCES

[1]. HVDC2000 - A new generation of HVDC converter stations, ABB pamphlet POW 0004.

[2]. CIGRE Study Committee No.33, "Application guide for insulation coordination and protection of HVDC converter stations", 33.83 (SC 03-21, WD).

Chapter 13

Case Studies Of AC-DC System Interactions

13.1 INTRODUCTION

The transient/dynamic behavior of integrated ac-dc systems is closely related to the characteristics of the inter-connected ac-dc systems and their controllers. DC controllers are fast-acting and respond within tens of milliseconds and influence the transient behavior. AC generators/machines are slow-acting and respond within hundreds of milliseconds and influence the dynamic conditions of the power system. Predicting the transient/dynamic behavior of integrated ac-dc systems therefore requires a suitable tool such as a powerful simulator to study these.

In this chapter, three typical simulator-based case studies are presented to demonstrate the capabilities of simulators - with their full representation of controllers and detailed system models - to highlight some particular system aspects of multi-terminal HVDC systems and multiple infeed HVDC systems.

13.2 AC-DC SYSTEM INTER-ACTIONS

This section provides some insight into two aspects which play a role in the inter-actions between ac-dc systems.

13.2.1 System Aspects

The three major elements concerned with inter-actions in ac-dc systems (Figure 13-1) are the dc converter, the ac system, and the dc system.

- The dc converter acts as a source of harmonics; current harmonics on the ac side of the converter, and voltage harmonics on the dc side. It also acts as a modulator of harmonics between the ac and dc sides of the converter. Converters also act as modulators of harmonics between its ac and dc sides. For instance, a 120 Hz current on the ac side will be converted into 60 Hz and/or 180 Hz on the dc side. Similarly a 60 Hz voltage on the dc side will be reflected as dc and/or 120 Hz on the ac side. The dc converter is also a power amplifier and acts a sink of reactive power for both rectifier/inverter modes of operation,
- The ac system acts as a source/sink of harmonics, a source/sink of active/reactive power and as a variable frequency-dependent impedance, and
- The dc system acts as a source/sink of harmonics, a source/sink of active power and as a variable frequency-dependent impedance.

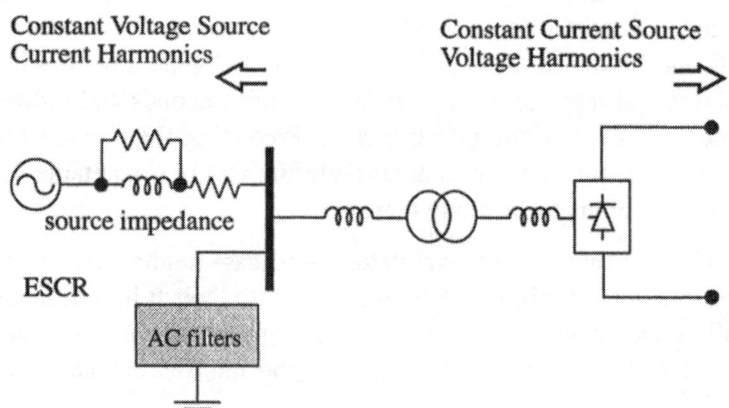

Figure 13-1: AC-DC system inter-actions

The factors which come into play for the system inter-actions are:

- AC-DC system strength,
- DC power transfer limits,
- Control and protection aspects for the ac-dc system,

- Reactive power co-ordination,
- Instabilities and harmonic transfer capabilities,
- Recovery of the system from faults,
- Temporary over-voltages and insulation co-ordination,
- Zero or low inertia systems,
- Harmonic filters and resonance phenomena,
- Non-linear elements (i.e. transformers, arrestors),
- Dynamic loads, and
- Others.

13.2.2 DC Controller Aspects

The operating point for a MTDC system is defined by the intersecting point of the static V_d - I_d characteristics of the terminals forming the dc system. Each linear part of the V_d - I_d characteristic defines one operating mode. Such operating modes are utilized in dc systems for reasons of security and/or efficiency of power transmission. Mode shifts (defined as the movement of the intersecting point from one linear part of the static V_d - I_d characteristic to another) are an integral part of system behavior and can delay system recovery from perturbations. However, when planned for, they can be beneficial to reduce the impact of system disturbances.

In the case of a MTDC system, the number of operating modes is high; for a n-terminal system there exist 2n principal operating modes. Consequently, the range of operation before a mode shift occurs is relatively limited and mode shifts occur more frequently. The choice of current margin is important for MTDC systems with terminals having a large disparity in their ratings. The magnitude of the current margin should be selected to ensure that the smallest terminal is not subjected to large overcurrents due to a mode shift.

13.3 MULTI-TERMINAL HVDC SYSTEMS [1,2,3]

Reference [1] describes a now classical simulator study of a four-terminal HVDC system with a small parallel tap (Figure 13-2). This study represents one of the largest systems ever to be mounted on a simulator, and has become a *defacto* benchmark for such studies.

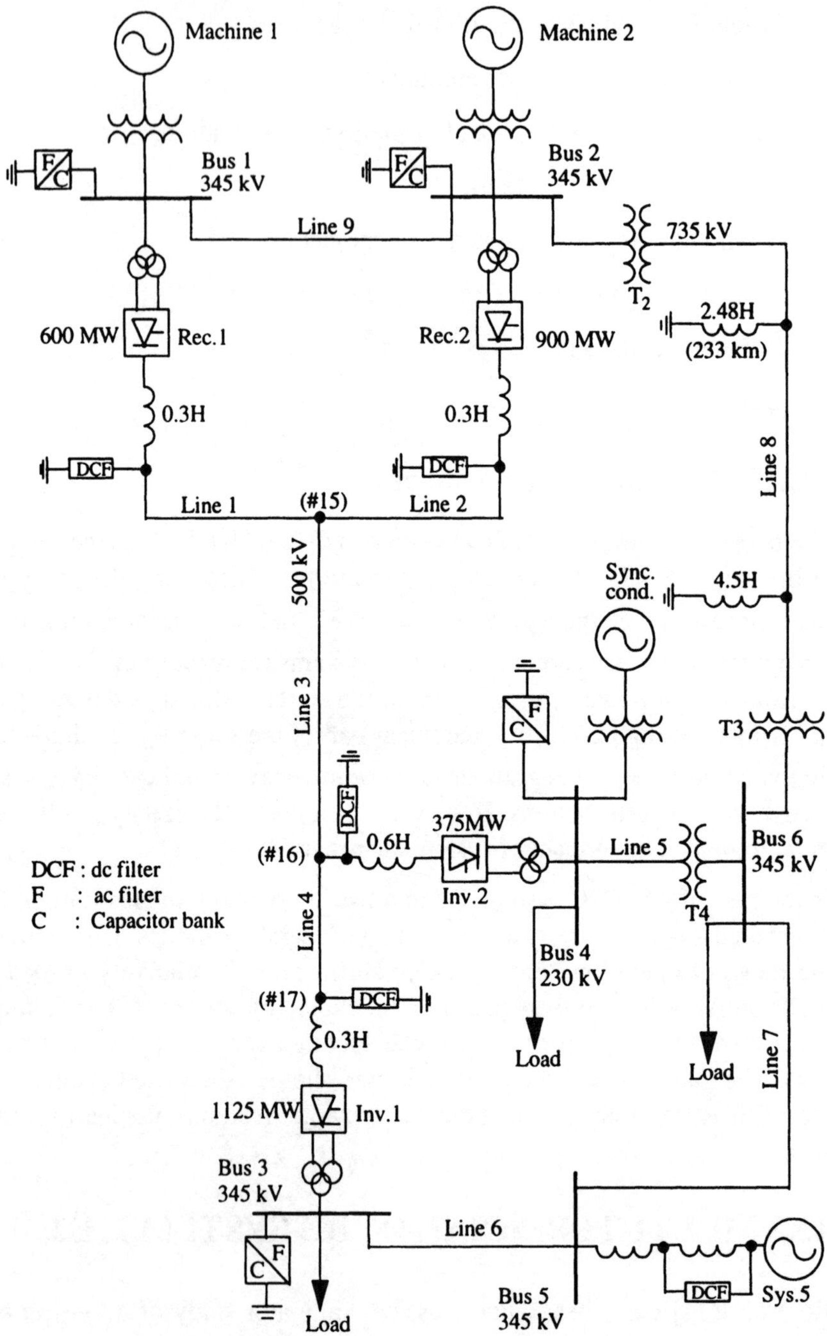

Figure 13-2: Four terminal MTDC System [1]

The system modelled is a 500 kV, 1500 MW four-terminal HVDC system with a small parallel tap which is inter-connected to an ac network with weak ac systems at the four dc terminals. The small tap is rated at 25% of the rating of the HVDC system and the effective short circuit ratios at the dc terminals are in the ratio of 2.1 to 3.0. The control strategy employed for the system was the Current Margin method.

The static V_d - I_d characteristics of the MTDC system are shown in Figure 13-3. The nominal operating point is point 1, with Rectifier 2 in voltage control (VC) and the remaining terminals in current control (CC).

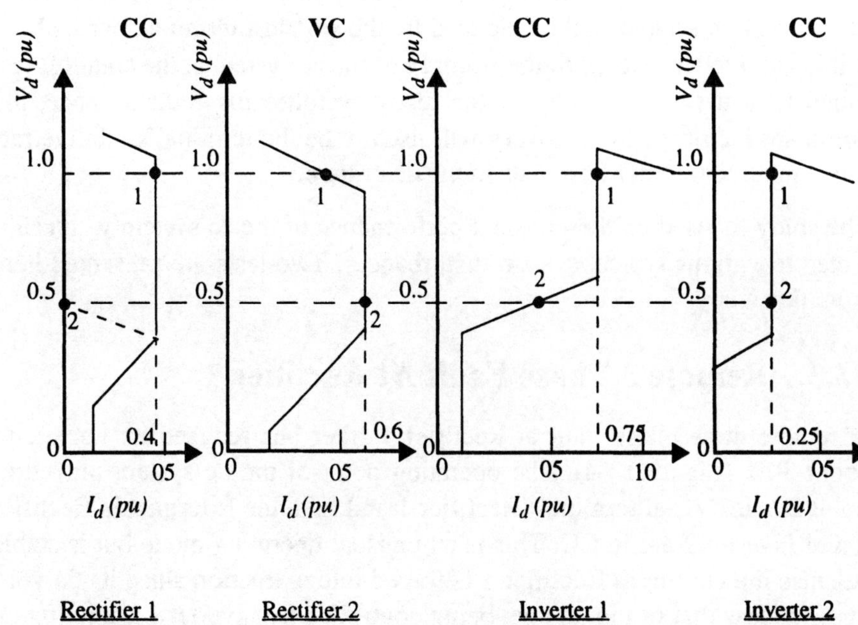

Figure 13-3: Static V_d-I_d characteristics of the MTDC system [1]

The relative rating of the four terminals are indicated in Table 13-1, as well as the Effective Short Circuit Ratio (ESCR) of their respective ac systems.

Table 13-1: Relative rating of the four terminals

	DC System	Rec1	Rec2	Inv1	Inv2
Rating (MW)	1500	600	900	1125	375
Rating (%)	100	40	60	75	25
ESCR	-	3.0	3.0	2.8	2.1

In the case of MTDC systems with unequally rated terminals the recovery of the total dc system will be dictated by the smallest terminal, particularly if it is an inverter. Although the strength of the ac system at the smallest terminal will affect the quality of the recovery following a disturbance, the dominant factor for the recovery will usually be the terminal's relative rating compared to other terminals in the dc system.

The study focused on the transient performance of the dc system when subjected to various typical system disturbances. Two tests are presented here from this study:

13.3.1 Remote 3 Phase Fault At Rectifier 1

A remote three phase fault at Rectifier 1 filter bus reduced its voltage to about 30% (Figure 13-4). The operating point of the dc system moved to point 2, with VC at terminals Rectifier 1 and Inverter 1; terminals Rectifier 2 and Inverter 2 are in CC. This is an unusual operating mode but is stable because the current in Rectifier 1 is forced into extinction since its dc voltage is below that of the dc line, being controlled by Inverter 1 operating on its alpha-min-limit-in-inverter-mode.

In the post-fault period, due to the dynamics of the ac-dc system a number of dynamic mode shifts between the two rectifiers occur. This is clearly observed in the trace of "Alpha Rectifier 2", which goes in and out of alpha-min operation. Such multiple mode shifts are a result of the breakpoints of the *Vd - Id* characteristic being relatively close to each other causing interaction between control and system dynamics. Such interactions in MTDC systems can lead to further disturbances and instability, and should be carefully assessed.

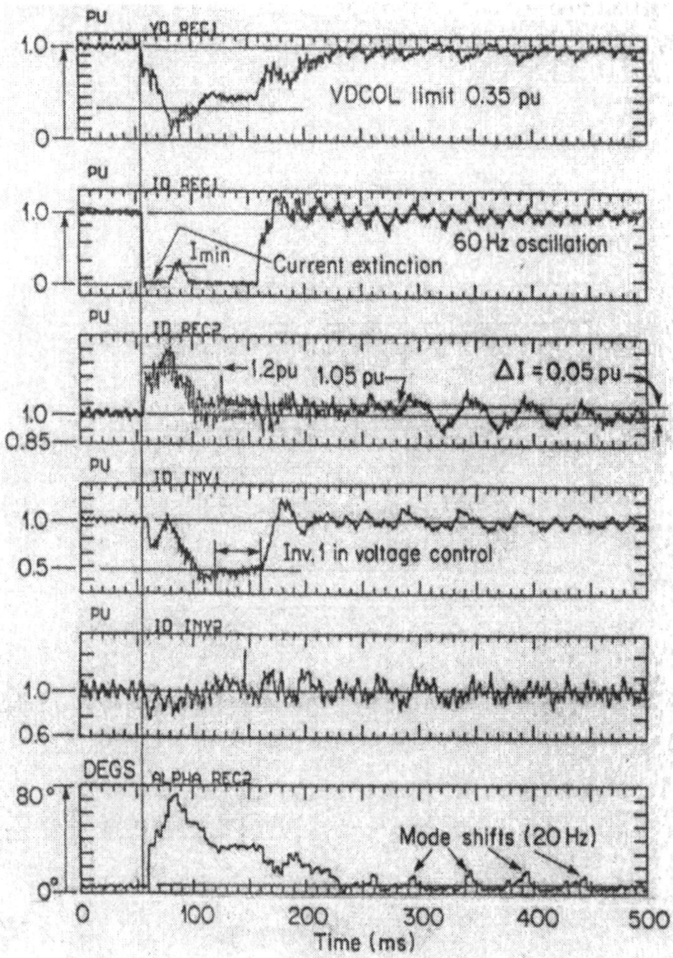

Figure 13-4: Remote 3 phase fault at Rectifier 1

13.3.2 Commutation Failure At The Small Inverter 2

The case of a commutation failure at the smaller inverter is shown in Figure 13-5. A commutation failure at the smaller inverter (Inverter 2) will cause the total dc line current to be diverted into Inverter 2 i.e. a steady state current value of 4 times its nominal value; with the effects of line capacitance discharging into the fault, a transient peak current value in the region of 6-8 times is feasible. This peak current can be limited partly by choice of a larger smoothing reactor at the smaller terminal. This magnitude of fault current is difficult for the ac commutation voltage to control (even with the support provided by a strong ac system).

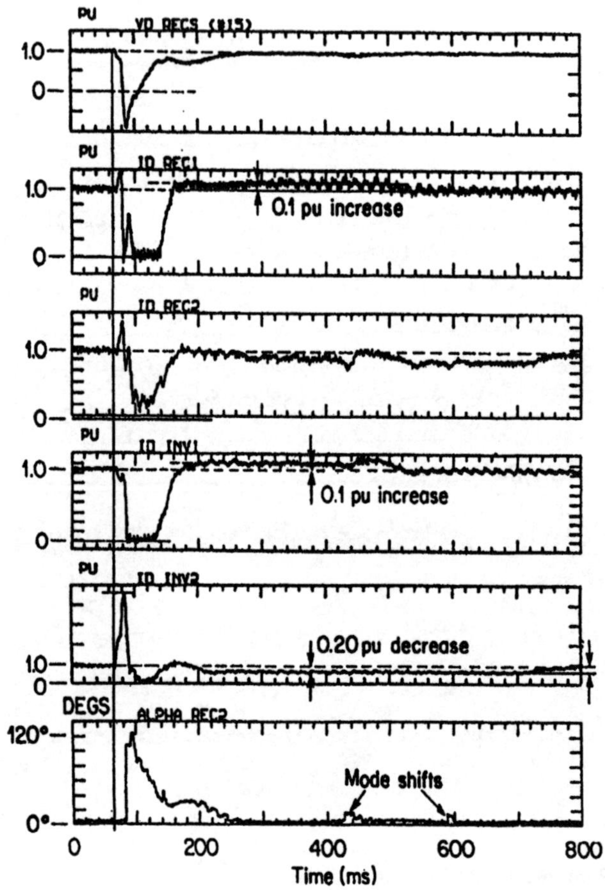

Figure 13-5: Commutation failure at Inverter 2

The recovery of the system can be obtained by utilizing the capability of the rectifiers to Force-Retard (FR) their firing angles (alpha) into the inverter region of operation i.e. 145 degrees or so. The application of the FR can occur without the need of telecommunications since the collapse of the dc voltage at the rectifier-end can be utilized as the signal to initiate the protection. After a pre-determined time the FR can be removed and the converter firing angles returned to normal at a controlled speed. However, the danger from a subsequent commutation failure at the smaller inverter persists under such conditions, and special control procedures may be necessary, especially with the weak ac system present at Inverter 2.

The danger of a subsequent commutation failure is highest when the current at the smaller inverter is nearing its nominal value, especially if an overshoot occurs due to, say, improperly tuned control parameters. Furthermore, the ac voltage is coincidentally at a low value due to sudden reactive power demand from the converter (Figure 13-5). This danger can be safely minimized by reducing the ordered current at this terminal and avoiding any transient overshoot of the actual current above this value. Optimization of the control parameters means having the smaller inverter react as fast as possible with minimal overshoot. The current regulator of the voltage controlling terminal is biased off such that it absorbs any leftover current. The means to reduce the current at the smaller inverter can be incorporated into its dynamic Voltage dependent Current order Limit (VDCOL) characteristic having special complex ramps (Figure 13-6). The utilization of these complex ramps is justified because they optimize the reactive power capability of the receiving-end ac system to support the ac voltage whilst accepting the maximum dynamic energy (MW-sec).

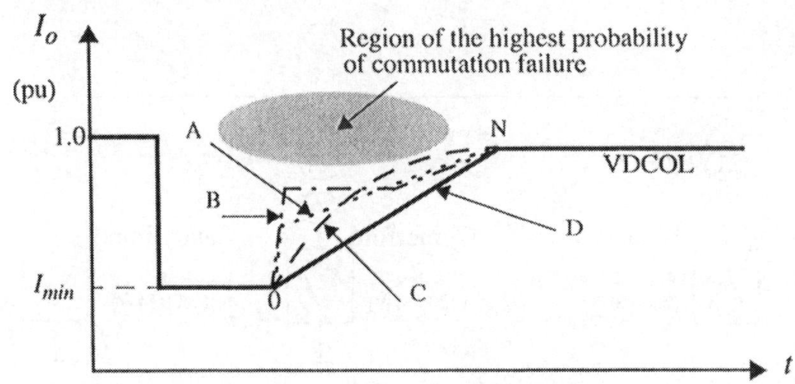

Figure 13-6: Dynamic VDCOL with complex ramps

Examples of ramps are shown in Figure 13-6. Ramp D is a linear ramp starting at minimum current order (point O) and ending at reaching nominal current order (point N). Ramp C is a double linear ramp and has the same starting point and ending points; however, it has a better MW-sec transmission capability when compared to Ramp D. Similarly, Ramp B with its delayed double ramp offers a better MW-sec transmission than any other ramp. One of the simplest yet most effective "ramps" is the exponential Ramp A. In practice, Ramp B was found to offer the best performance in MW-sec transmission.

The findings from this study showed that:

1. The recovery of the system shows a marked oscillation at about 20 Hz due to mode shifts occurring at the rectifiers. This is a kind of limit cycle and can generate a low frequency oscillation over an extended period of time. The presence of generators near the rectifiers is of concern since SSR type problems may arise. This test shows the negative influence of mode shifts in the system.

2. The second case shows the recovery of the system following a commutation failure at the small Inverter 2. This causes a severe over-current at the Inverter 2. Again mode shifts can be observed at Rectifier 2.

The negative impact of mode shifts was shown in the test case above. However, mode shifts can be useful too, as was shown in a study of a three-terminal MTDC system (Figure 13-7) [3].

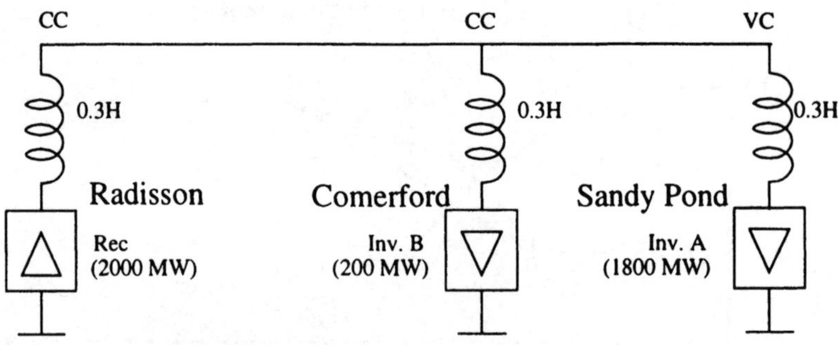

Figure 13-7: Three terminal MTDC system

The system characteristics of the small inverter B at Comerford, cause it to suffer from a second harmonic resonance condition, and the dc line is known to be resonant at 60 Hz. The case is presented from this study where the impact of a 3 phase fault at the Radisson (Rectifier) causes a severe injection of second harmonic voltage/currents into the dc system. With the presence of a second harmonic ac filter (in itself, an expensive option) at Comerford, the system normally suffers from a commutation failure at recovery time and delays the recovery period considerably. Without this second harmonic filter at Comerford, the system manages to recover inspite of the resonance condition (Figure 13-8a). However, changing the alpha-min limit setting at the Comerford from a low to a high value causes the dc system to recover even with a second harmonic filter present (Figure 13-8b). The application of this mode shift is, therefore, adapted to be a useful feature. A further improvement of the condition results if a delayed dynamic VDCL is employed [3].

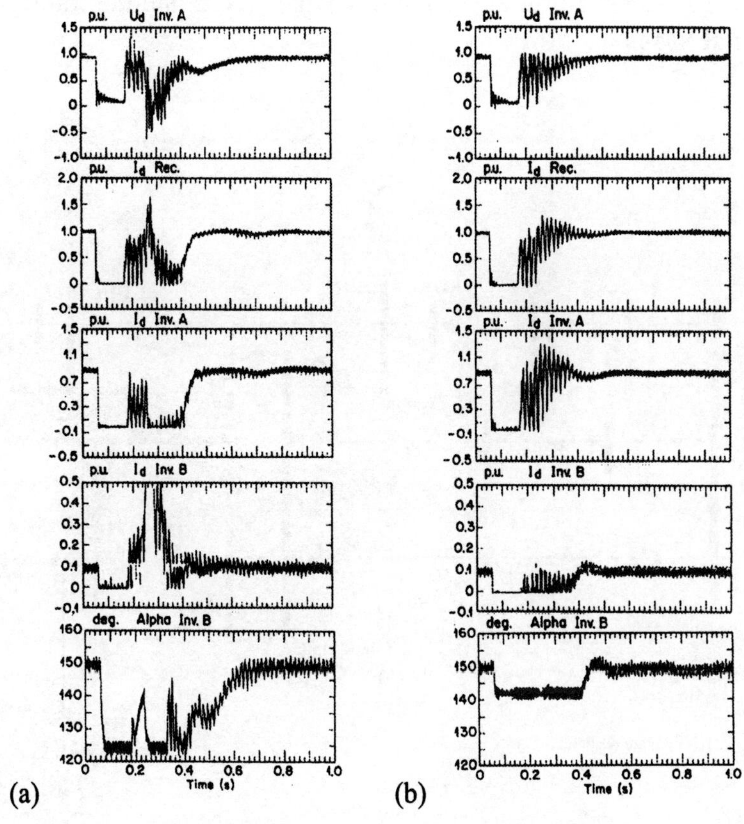

Figure 13-8: Recovery of a three terminal MTDC system [3]

13.4 HARMONIC INTER-ACTIONS AT CHANDRA-PUR HVDC STATION [6]

In the case of two asynchronous ac systems inter-connected through a dc link, another aspect of converter modulation is observed. Usually a beat frequency (i.e. the difference of the two ac frequencies) is observed on the dc link. In the case of two ac frequencies being approximately the same but slightly phase shifted, a very low frequency (usually under 1 Hz) oscillation is observed on the dc link. In the case of multiple infeed converters, this beat frequency will travel through from one dc system into the other.

The 1500 MW Chandrapur HVDC system (Figure 13-9) was studied on the simulator. The Chandrapur bus CHA4 is electrically very close to another HVDC 1000 MW BB at the bus WES4 which feeds into the Ramagundam system RAM4.

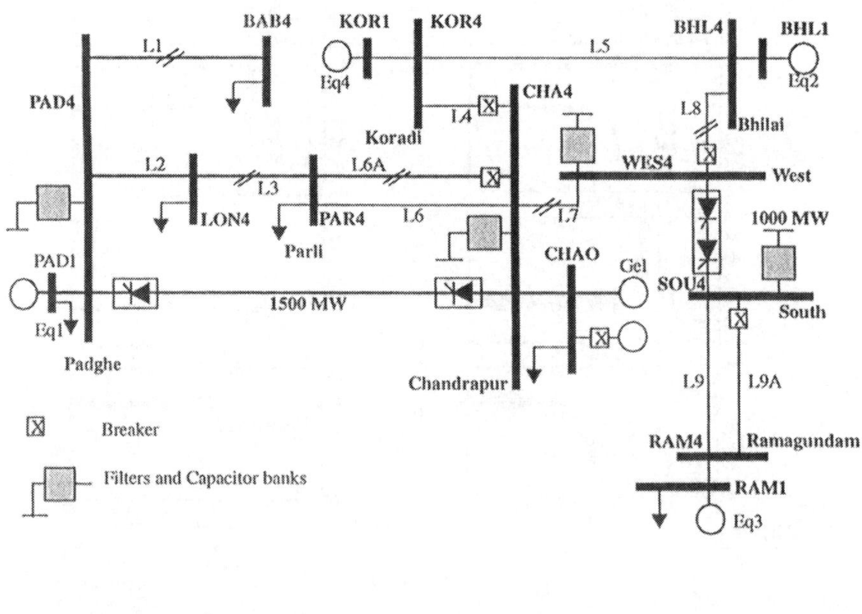

Figure 13-9: One line diagram of the Chandrapur-Padghe HVDC Inter-connection

In this particular case, the Ramagundam ac system is asynchronous from the Chandrapur ac system. However, since it is not possible to maintain the two systems at exactly the same frequency, a small (typically less than 1 Hz) oscillation at the difference (beat) frequency of the two systems i.e. at (Chandrapur frequency - Ramagundam frequency), is observed on the BB dc system. Furthermore, this very low frequency beating phenomena is also modulated by the Chandrapur converter into the bipole dc system.

As a demonstration of this phenomena, a 15 Hz signal was injected into the Chandrapur current order. Figure 13-10 shows the impact of this on the signals Ud, Id and firing angles of the Chandrapur and South converters.

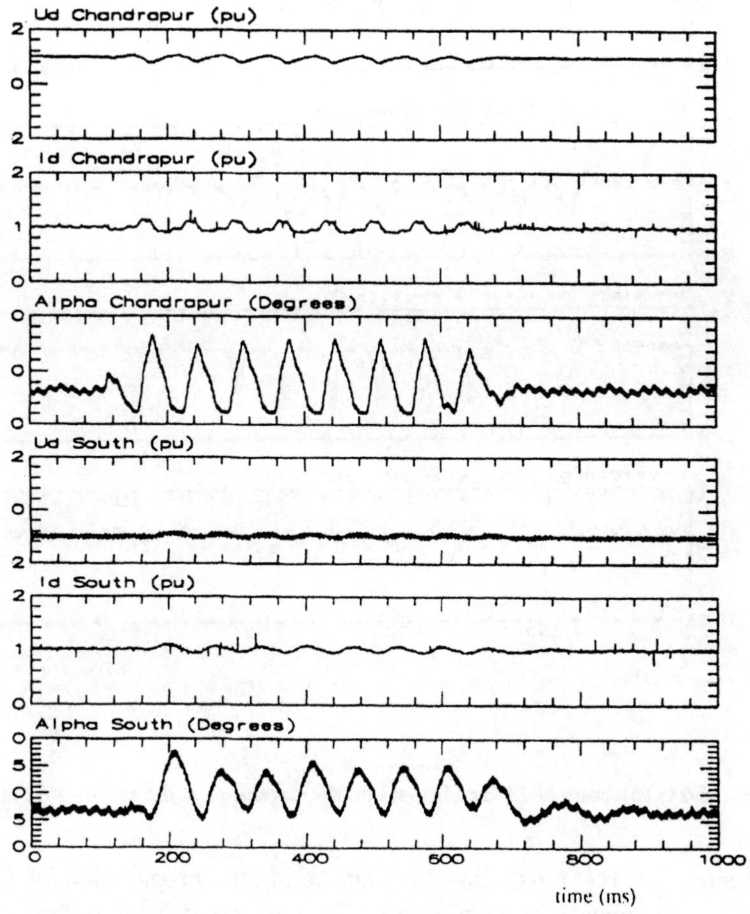

Figure 13-10: Injection of 15 Hz [6]

The impact of these is also seen on the envelopes of the three phase ac voltages at the four converter buses (Figure 13-11). Thus, any low frequency modulation effect from one dc system is propagated onto the other dc system. This modulation effect was observed for frequencies from 1 Hz upto about 120 Hz, after which the dc controllers were able to filter out the impact.

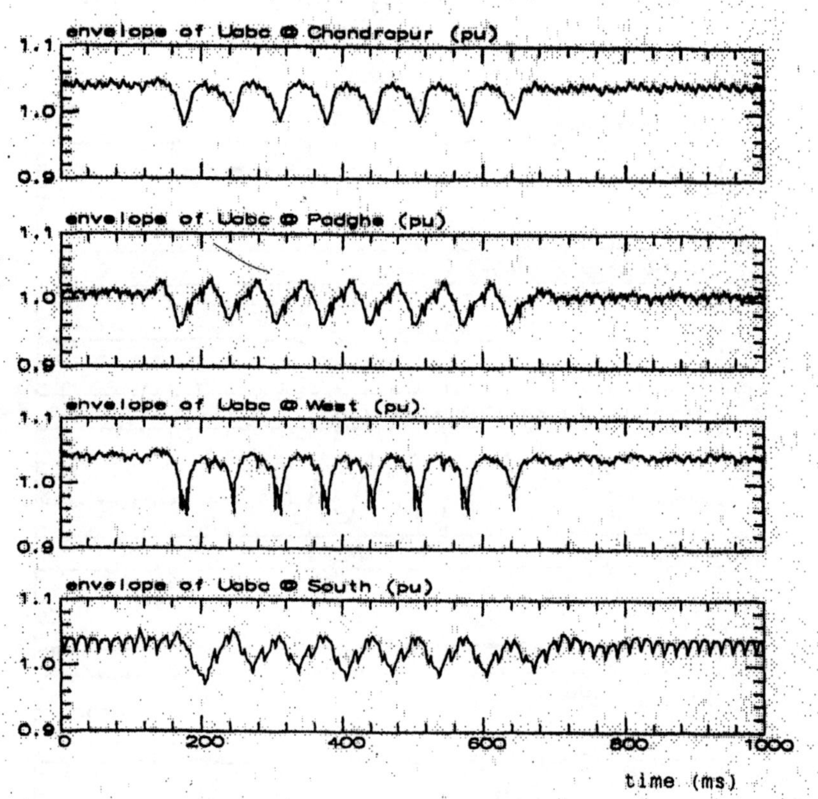

Figure 13-11: Injection of 15 Hz - Impact on the envelopes of the ac bus voltages [6]

One simple way of reducing the impact of this propagation of low frequency harmonics is to employ a realistically sized dc smoothing reactor. For the simulator study, this smoothing reactor in the BB link was 30 mH, which although small is representative of today's market place. Alternatively, using a control mode other than current control at the rectifier (such as constant-alpha mode) may also aid in decoupling this propagation effect.

The impact of propagation of such low frequency oscillations would have to be studied for particular system conditions. It is important to note that low frequency oscillations cannot be decoupled by the converter systems.

13.5 CONCLUSIONS

Due to the complex inter-actions that take place between integrated ac-dc systems, it is presently essential to use a sophisticated simulator to study their behavior. The equivalent simulator model for study must be carefully derived from the actual system to be designed/studied, in keeping with the overall objectives of the study. The representation of the ac-dc systems should be in adequate detail and the study should preferably be carried out with a faithful replica of the real controls of the dc systems.

Due to the economic cost of a simulator study and its potential benefits to the overall system design and performance, the study should be well-planned and professionally executed by well-trained personnel. From the case studies presented in this paper and the experience gained, it can be said that the insight derived from such studies is usually well worth the investment.

13.6 ACKNOWLEDGEMENT

This chapter is dedicated to my many colleagues at the Simulator Group, IREQ whose many contributions have made such studies possible.

13.7 REFERENCES

[1]. P. C. S. Krishnayya, S. Lefebvre, V. K. Sood and N. J. Balu. "Simulator study of multiterminal HVDC system with small parallel tap and weak ac systems". IEEE Trans. on Power Apparatus and Systems, October 1984, Vol. PAS-103, No. 10, pp 3125 -3132.

[2]. V. K. Sood, "DC Controls based Protection to Enhance Operation of MTDC Systems". MONTECH IEEE Conference, 29 Sept. - 13 Oct. 1986.

[3]. V. K. Sood, H. L. Nakra, B. Khodabakhchian and G. Scott, "Simulator Study of Hydro-Quebec MTDC Line from James Bay to New England", IEEE Trans. on Power Delivery, Vol. 3, No. 4, October 1988, pp 1880-1886.

[4]. L. X. Bui, V. K. Sood and S. Laurin, "Dynamic interactions between HVDC systems connected to ac buses in close proximity", IEEE Trans on Power Delivery, January 1991, Vol 6, No 1.

[5]. A. Gole, V. K. Sood, "Modeling of Static Compensators and dc Converters for Control studies using Digital Simulation Programs", Canadian Electrical Association Spring Meeting, March 1990, Montreal.

[6]. V.K.Sood, "Simulator study of the Chandrapur - Padghe HVDC link", Report No. IREQ-93-028, February 1993.

[7]. J.Gagnon, A.Venne, "Operating Experience with the Chateauquay HVDC Tie", Canadian Electrical Association, 1990.

Chapter 14

Simulators For Analyzes Of Power System Phenomena

14.1 INTRODUCTION

Modelling of power system transient and dynamic phenomena is essential to study their impact on the network design and operation. Although modelling and simulation can be expensive, they can be potentially cost effective by the savings resulting from the insight they provide in the design and operation of power systems. Transient Network Analyzers (TNAs) were used in the past for power system studies. Essentially these TNAs were scaled-down analog versions of the power system. Today, three types of simulation tools are in use for the analysis of power systems:

- Hybrid simulator (i.e. IREQ simulator)
- Off-line digital simulation packages (i.e. EMTP and its derivatives),
- Real-time digital simulators (i.e. HYPERSIM from Hydro-Quebec and RTDS™ from RTDS Technologies Inc.).

Technical requirements for the simulator should cover the whole spectrum of:

- Accuracy/confidence level of the model,
- Functionality/flexibility of the model,
- Rapidity/capacity of the tool to perform a complex task, and
- Repetitive mode in real-time.

The simulator should provide:

- A 3-phase model to reproduce transient phenomena within the frequency range 0-2500 Hz,
- A facility to interface with actual system control and protection systems,
- Detailed models of system components, e.g. dynamic loads,
- Sophisticated Data Acquisition System,
- On-line Data processing facility,
- Computer controlled configuration changes, test execution and results management software, and
- Post-Processing of Data, Graphic Display and Report editing features.

14.2 THE IREQ HYBRID SIMULATOR [4-7]

First commissioned in the early 1970s, this famous simulator (Figure 14-1) has undergone continuous upgrading and evolution since its conception to keep pace with technological advances. This simulator was used extensively to study numerous HVDC and AC systems from all around the world i.e. the Manitoba Nelson River HVDC system, Itaipu Scheme in Brazil, Chateauguay HVDC BB in Quebec (Canada), the Intermountain Power Project (IPP) in the USA, the James Bay Multi-terminal DC system in Quebec (Canada), the Chandrapur-Padghe HVDC system in India etc.

The simulator is a small-scale, 3-phase replica of an actual power system. It uses well-proven and reliable electronic (analog and digital) technologies for the simulation of power system elements by lumped resistor, inductor and capacitor (RLC) components. Its rated operating voltage of 100 V phase to phase and maximum short circuit current of 5 A were selected to minimize the losses and cost in the simulation of power systems in the EHV and UHV range.

The simulator operates in real-time at the nominal power system frequency of either 50 or 60 Hz. Real-time operation permits the verification and optimization of real controls and protection equipment which can be interfaced to the simulator. The performance of the controllers and protection equipment can be verified and optimized under realistic operating conditions. Furthermore, operating conditions that are very severe but probable and contingencies that are often difficult but practically impossible to duplicate in the field can also be considered.

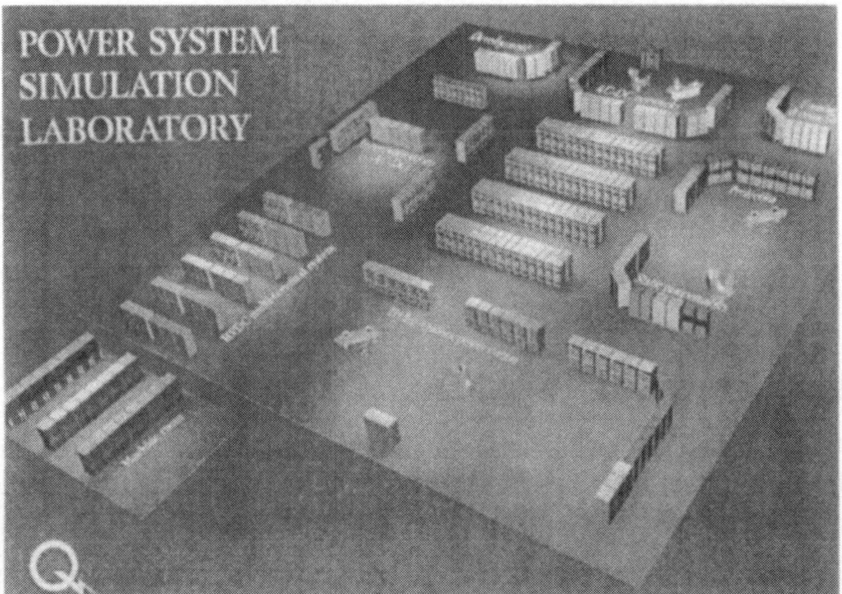

Figure 14-1: An artist's view of the IREQ Power System Simulator [7]

A major advantage of these types of simulator is the speed of test execution i.e. in real-time. Thus, the simulation of a phenomena that lasts 1 second takes 1 second unlike simulation in deferred (off-line) time where it may take many minutes or even hours. This feature permits detailed optimization studies to be undertaken where rapid interaction with the user is necessary to investigate complex phenomena and trouble-shoot controllers. Often such studies involve a "what if" type of scenario where the next test may depend on the results of the previous test. Therefore, it is imperative to minimize the time between tests.

The key features of the IREQ simulator are:

- Real-time simulation,
- Hybrid technology that includes passive, electronic (analog and digital) models and real-time digital models,
- Large number of components for simulation of large networks,
- Many versions of FACTs (SVC, HVDC) controllers,
- Inter-connection capability with commercial control and protection systems,
- Computerized systems for test automation, data recording, on-line analysis, post-processing and report editing, and
- An integrated database system for fast storage and recall of results and input data.

14.2.1 Modelling Techniques

Passive analog RLC components are used to model transmission lines (Figure 14-2), transformers (with saturation), shunt reactors, constant impedance loads, series and shunt capacitors and filters etc.

Electronic components are used to simulate arrestors, circuit breakers and thyristor or other static valves. Negative resistors are used to compensate for the disproportionate copper losses involved with the scaled down physical models.

Real-time digital models are used to simulate components such as controllers (Figure 14-3), synchronous machines with their regulators [8], FACTs (SVC, HVDC etc.) controllers and dynamic loads.

Computer controlled electronic switches are used for modelling circuit breakers with their closing resistors and zero current breaking characteristics. The automated system permits the execution of complex sequences that can be randomly controlled for statistical and optimization studies.

System voltages are scaled down to 100 V phase to phase on the simulator. An impedance scaling factor is selected to keep the fault current below 5 A. The scaling factor makes it possible to study all practical phenomena within a dynamic range of 10 - 50,000 A. The frequency bandwidth of simulator components is limited to about 2.5 kHz (note that this bandwidth is also a function of the complexity of the models used).

Figure 14-2: Photo shows a row of RLC components connected to model transmission lines

Figure 14-3: Photo shows a set of HVDC digital controllers, supplied by ABB, for the Hydro-Quebec multi-terminal dc system at the IREQ simulator

Monitoring equipment allows the monitoring of system phenomena to well above 20 kHz. The computerized recording equipment for each test bay is comprised of one A/D converter with 128, 13 bit analog multiplexed input channels and 64 digital multiplexed input channels (Figure 14-4) for a total of 192 input channels. The transfer rate is 1 mega-samples per second. A 12 Mb memory is dedicated to each A/D and since it is directly connected to the main computer, the data transfer time is eliminated.

Computer controlled network inter-connection panels are used extensively in each test area. Each model used in a study is inter-connected via these panels. The network status is therefore automatically monitored by computer such that the network status conforms to the study database information for each test. The unique fingerprint of each test is stored and can be easily verified for conformation of network integrity. This technique has been found very practical with the simulation of large complex networks.

A waveform playback system is in use for protection system verification with previously recorded waveforms recorded from either the simulator, EMTP simulation or field data. This playback system is very useful for checking corrections made to protection systems after the original simulation set-up has been dismantled [10].

14.3 OFF-LINE DIGITAL SIMULATION PACKAGES

14.3.1 EMTP

The Electro-Magnetics Transients Program (EMTP) was developed by H. Dommel [15,16] at the Bonneville Power Administration (BPA). Its fundamental simplicity resulted from the representation of inductor (L) and capacitor (C) elements by resistor (R) elements with initial (history) conditions. It used the trapezoidal integration method which was both simple and rapid in the solution of algebraic equations which could be represented by difference equations. The network equations are obtained using the Nodal Analysis method. EMTP uses sparsity in the inversion of matrices and the use of LU factorization techniques made the solution of large systems relatively rapid. The user-specified time-step of integration was kept constant throughout the simulation. The availability of a sub-routine for the Transient Analysis of Control Systems (TACS) [12] enhanced its capabilities and allowed the use of control systems to be added to the simulation. This sub-routine was later replaced with another more powerful version called MODELS [13,14].

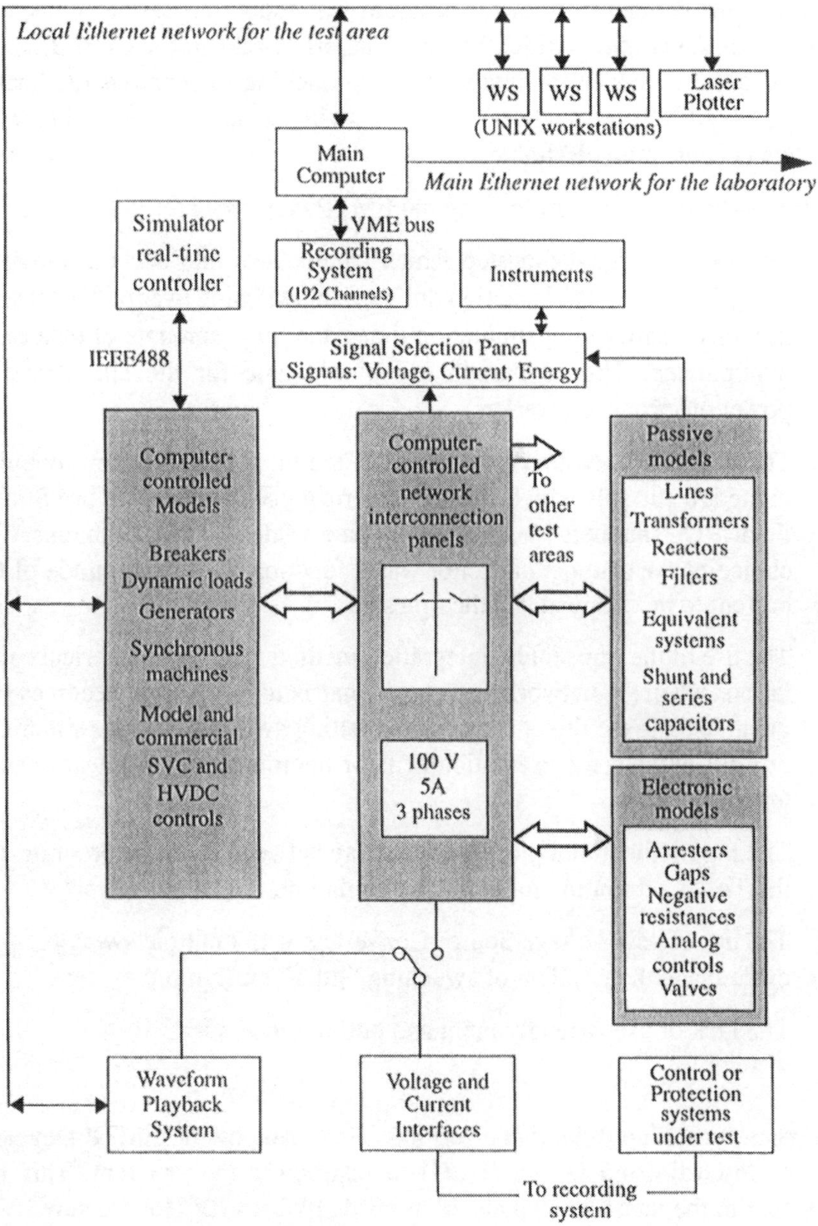

Figure 14-4: A typical setup arrangement on the IREQ simulator

EMTP rapidly became a industry standard analysis tool for power systems and is widely used. The program has had a checkered history and numerous variants have appeared, including the Alternative Transients Program (ATP) version which could be used on a personal computer.

Initially, the development of the program was supported by the Bonneville Power Administration (BPA). Some of the drawbacks in the capabilities of EMTP became more pronounced as the modelling of Flexible AC Transmission (FACTS) with power electronic switches and voltage source converters became more desirable.

Some of the drawbacks of the original EMTP version were:

1. The use of a fixed timestep which cannot take into account the relatively long periods of inaction during non-switching events. This results in unnecessarily long simulation times and huge amounts of data to be manipulated. This is particularly problematic for the simulation of power electronic converters.
2. The use of a fixed timestep results in the modelled switches chopping inductive currents which causes numerical oscillations. The use of artificial RC "snubbers" helped to alleviate some of these problems. The choice of the snubber capacitor was a function of the magnitude of the current to be chopped and the timestep.
3. The use of the trapezoidal integration method results in numerical oscillations when the network admittance matrix to be inverted becomes singular. This is the direct result of modelling switches as truly either ON or OFF without representation of their intermediate non-linear characteristics.
4. The requirement of a one timestep delay between the main program and the TACS subroutine for controls simulation.
5. The use of new Voltage Source Converters with multiple switchings per cycle made the problem of switching "jitter" much more evident.
6. The lack of user-friendly input and output processors.

In recent years, considerable effort has been made by the EMTP Development Co-Ordination Group (DCG) to restructure the program. This has resulted in the latest version called the EMTPWorks RV (Restructured Version). The entire code of the program has been re-written and graphical input and output processors have been added. A sample of the graphical input file is shown in Figure 14-5 and a sample of the output file is shown in Figure 14-6.

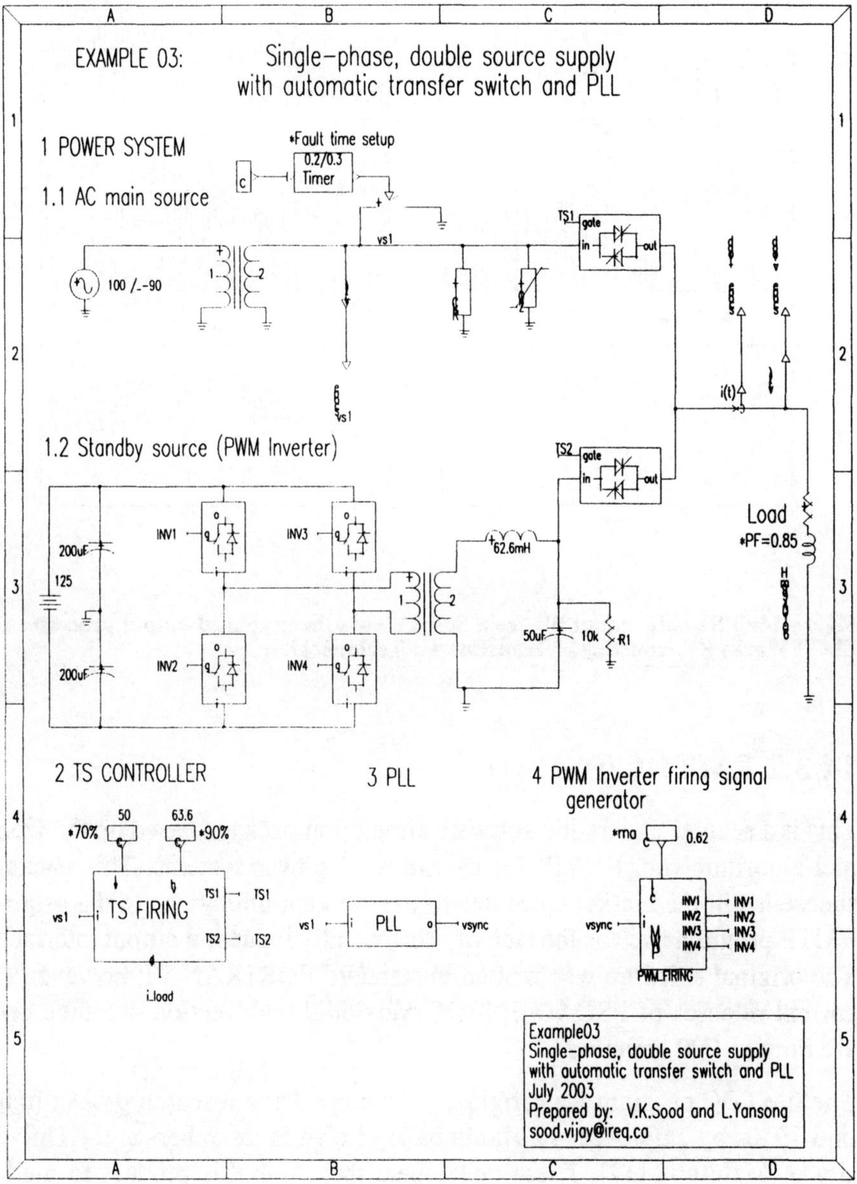

Figure 14-5: Sample graphical input file for the latest version of EMTPWorks RV (courtesy of EMTPWorks RV)

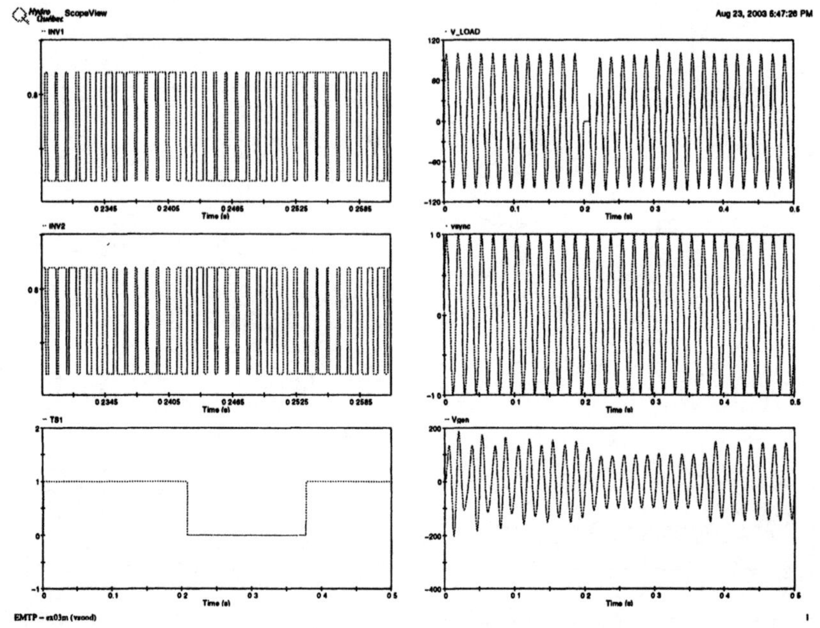

Figure 14-6: Sample output file from ScopeView - the graphical output processor for EMTPWorks RV (courtesy of TransÉnergie Technologies))

14.3.2 EMTDC/PSCAD

This is a second-generation software simulation package based on the Dommel algorithm (i.e. EMTP) for the study of power systems. This package succeeded in the market place due to the inherent limitations in the original EMTP program such as the lack of user-friendly input and output interfaces. The original software was written in standard FORTRAN 77; however, the current releases of PSCAD/EMTDC Version 3 and Version 4 source code are Fortran F90 compliant.

The EMTDC program was originally developed as a research project in the mid 1970s by employees of Manitoba Hydro and researchers at the University of Manitoba [17]. These developers then took this product to market under a new not-for-profit enterprise named Manitoba HVDC Research Centre (established in 1981) which is a research centre for Manitoba Hydro. In 1988, the Centre began development of a graphical user interface to enhance the EMTDC simulation software. The result was PSCAD with Version 2 being released in 1994 and Version 3 in 1999. Version 4 was released in the early part of 2002.

In recent years an interpolation algorithm has been included to permit the accurate modeling of power electronic switches. For the modelling of large power systems, the network can be split into smaller sub-networks connected by transmission lines (which effectively de-couples sub-networks by the time delay encountered between the sending and receiving ends). The solution of these individual sub-systems has been made efficient by the use of Gaussian Elimination sparse matrix techniques.

The PSCAD/EMTDC Master Library includes, but is not limited to, the following components:

- **Network Components**

  - Passive RLC components
  - Transformers (with saturation)
  - Frequency dependent transmission lines and cables
  - Synchronous and Induction machines
  - Breakers
  - Surge Arrestors, and
  - Sources

- **Control Blocks**

  - Derivative
  - Delay
  - Differential lag
  - Integrator
  - Limiter
  - Complex pole
  - Real pole
  - Lead lag
  - Ramp
  - Timer

- **Power Electronics**

  - Thyristors, Diodes and GTOs
  - 6 and 12 pulse HVDC converter bridges
  - SVC
  - STATCOM

- **Meters**
  - RMS volts and current (1 or 3 phase)
  - Real and Reactive power
  - Peak detector
  - Relative phase angle
  - Frequency

The software also has many motor models, new control functions, transmission line, and transformer models that are not listed above.

14.3.2.1 PSCAD Graphical User Interface (GUI)

This is a proprietary GUI (Figure 14-7) which can be used with EMTDC. This is a powerful GUI which facilitates the user to manage, prepare, run and analyze simulations.

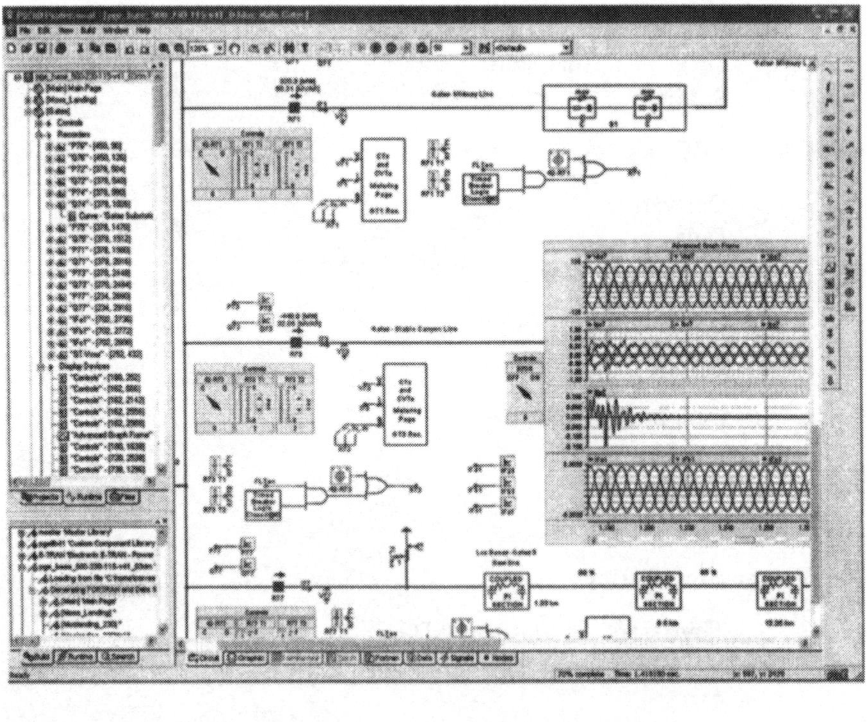

Figure 14-7: An example of the PSCAD GUI used with EMTDC

14.4 REAL-TIME DIGITAL SIMULATORS

A real-time digital simulator [1,2] is an important study tool for power system engineers involved with the analysis of complex power system interactions caused by the increasing use of fast-acting controllers and protection systems. The cost considerations for deployment of this tool are not insignificant and must take into consideration the infrastructure, operational, maintenance and upgrading costs.

Real-time simulators exist at the following organizations:

- Hydro-Quebec Research Institute (IREQ), Canada [4,5],
- Manitoba HVDC Research Center, Canada [3],
- Kansai Research Center, Japan,
- CEPEL (research center), Brazil,
- CESI (research center), Italy,
- CPRI (research center), India,
- EPRI in China,
- ABB, Siemens, GEC-ALSTOM, GE & other suppliers,
- EdF in France,
- Korea Electric Power Corporation, (with the largest RTDS installation),
- Others.

The main limitations of real-time simulators are:

- Limited number of models available at one site due to cost and physical constraints which limit the size of the power system that can be modelled,
- Single user capability of most installations limiting the availability of the simulator and the number of studies that can be performed within a limited time frame, and
- Operating costs (maintenance, operator training, model set-up, etc.).

It is noteworthy that the maintenance and operating costs and some of the limitations mentioned above are reducing all the time due to the digital approach. While real time digital simulators are not without cost, they are a fraction of the price of analog/hybrid systems (~1/10) and an even lesser fraction in size.

From the point of view of an utility, the needs for a real-time simulation capability are:

- Verification/Commissioning of actual control and protection systems in closed-loop mode,
- Feasibility studies and investigation of new technologies (i.e. FACTS devices),
- Investigation of complex systems with interactions possibilities, and
- Training of operators and engineers.

In many applications, it is necessary to compute the current and voltage wave forms at the same speed as they would occur in reality, independently of the size of the power system network being investigated by real-time simulators which can be made of:

- Physical passive component such as resistors, inductors, and capacitors,
- Electronic components such as operational amplifiers, zener diodes, thyristors etc., and
- Special Digital Signal Processors (DSPs) programmed to execute their program in real time.

Modern real-time simulators interface to these three technologies to maximize the operating flexibility and the precision of the simulation. However, due to the frequency bandwidth of physical components and to the speed limitation of actual digital processor technology, real-time simulators cannot be used to evaluate very fast transients with rise time smaller than 10 - 50 micro-seconds initiated by lightning strokes and arc breaking and re-ignition phenomena. With the exception of these very fast transients, all the other transient phenomena occurring during and after a system disturbance can be simulated in real-time including the reaction of fast and slow control and protection systems and electromechanical oscillations. This is a key feature necessary for the analysis of complex interactions between HVDC and FACTS controllers, line & equipment protection, automatic controls and the AC networks.

However, the transient stability performance of a very large power system cannot be investigated with today's real-time simulators due to the rather limited quantity of models available at one site. These studies are performed with off-line transient stability programs neglecting the simulation of all electromagnetic transients in order to decrease the simulation time and to allow the simulation of very large power systems.

14.4.1 Methodology

Most real-time digital simulators employ the well-known EMTP-based Nodal Analysis approach for simulating the network. This permits a global approach to the representation of RLC elements (Figure 14-8) as well as switching components. Each element is represented by a resistor in parallel with a current source; the current source represents a history term which contains the energy stored in a reactive element. Kirchoff's Nodal Law enables the building of a set of linear equations:

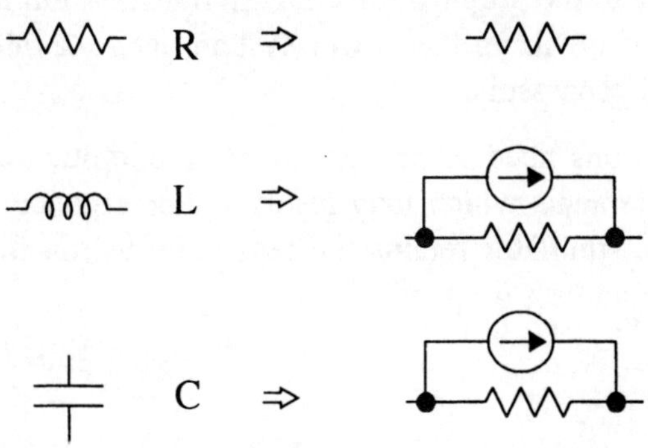

Figure 14-8: Nodal method of modelling RLC components

$$[G]\,[V] = [I] \qquad (14\text{-}1)$$

$$[V] = [G]^{-1}\,[I] \qquad (14\text{-}2)$$

By calculating the inverse of the admittance matrix [G], the solution to the system is obtained.

The trapezoidal integration method is usually utilized for efficient and rapid calculations, although care has to be taken to avoid singularity of the matrix [G] due to one branch having a very high impedance compared to the rest of the branches. This can lead to numerical oscillations; however, most power systems have enough damping in the circuit to avoid such problems.

A fixed time-step (typically 25-75 micro-seconds) is necessary with a trapezoidal integration method which is commonly employed. Since the current in a switching element in series with an inductive element will normally not coincide with a switching requirement, current chopping occurs and can lead to numerical oscillations. However, techniques are now available to include interpolation techniques to avoid numerical problems from occurring. The extra burden imposed on computation is not too severe with modern processors.

Jitter problems can arise when modelling control functions with very long time constants as compared to the time step. Use of powerful double precision processors is therefore recommended (i.e. DSPs and RISC processors).

Decoupling of the power system network is necessary to distribute the computational burden between parallel processors (Figure 14-9). The power system is decoupled due to the propagation delay in transmission lines, thus this technique is used to break up the power system network into smaller sub-systems for parallel processors.

Since real-time calculations need to be fast, iterative computational techniques for non-linear elements which may result in convergence problems are often not used. The simulator retains the feasibility to run in deferred time also for debugging and set-up purposes.

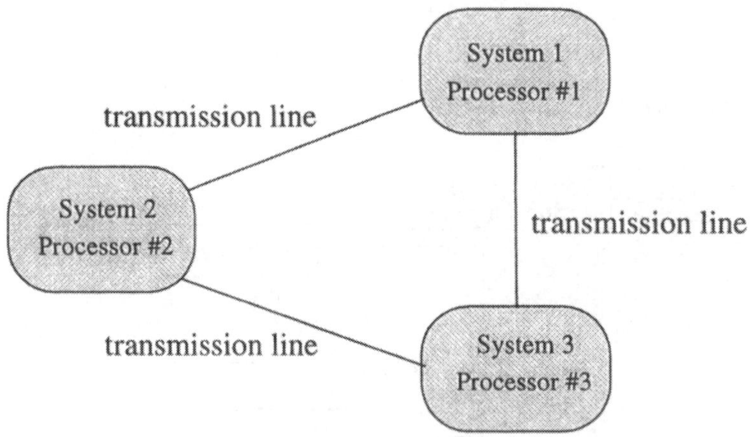

Figure 14-9: Technique to distribute tasks to parallel processors

14.4.2 Hardware Considerations

Since computational speed is of the essence, parallel processors are required. A time step of 50-75 micro-seconds is considered essential. The processor nodes must be capable of inter-nodal communication with at least 4 other nodes to minimize communications overheads. Usually each node will have an I/O capability to reduce the inter-nodal communications burden.

The availability of the computational nodes must, therefore, be a mature, stable and easily available technology. Furthermore, due to the coding requirements, portability to other machines is usually a consideration.

14.4.3 Software Considerations

Due to code portability requirements, a high level language such as C/C^{++} is commonly employed. Furthermore, Object Oriented Methods (OOM) (Figure 14-10) are used for the flexibility, modularity, expand ability and maintenance purposes of the code. Another benefit of OOM is the ability to employ inheritance relationships to reduce the number of object classes.

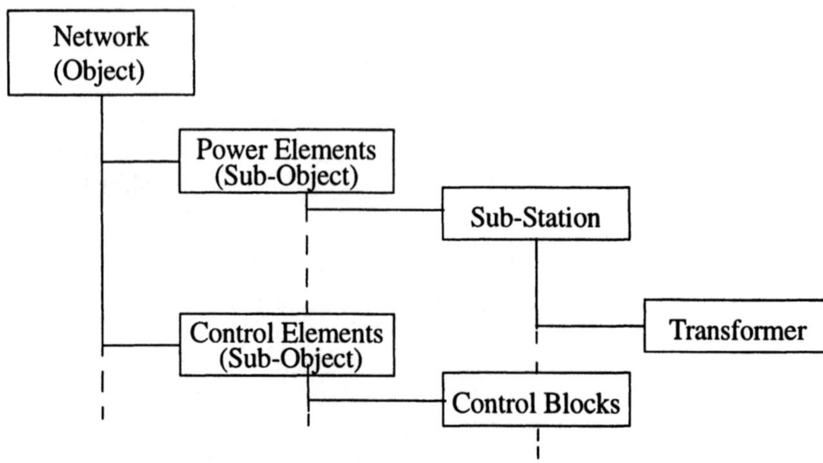

Figure 14-10: Object Oriented Methodology for software

14.4.4 Graphical User Interface (GUI)

Since this is the main interface with the user, its friendliness and ease-of-use is of primary concern. Three separate functions of the GUI are identifiable:

1. Editing Mode

The usual tendency is to employ a graphical window for drawing the power system network from a library (palette) of elements. The user usually has the facility to develop in-house icons and elements which can be added to this library. The icons can be manipulated for easily entering the data parameters.

2. Control Mode

This mode enables the user to manipulate the simulation and control it on-line. Features such as ending and re-starting a simulation with modified parameters are common. Added facility can be the operation of multiple case studies for performing sensitivity analysis or optimization type studies.

3. Data Analysis Mode

In this mode, the vast amount of data that is acquired during a simulation is manipulated for fast analysis and visualization. The use of sophisticated data processing techniques is a requirement.

14.4.5 Validation Of Real-time Digital Simulators

Validation is both necessary and essential to any analytical tool. Validation can be performed with results from field tests, other simulators/TNA and deferred-time digital packages. The use of benchmark tests i.e. the CIGRE Benchmark HVDC system has become standardized now.

14.4.6 Hardware Implementations

There are two well known versions available. The first one is from the RTDS Technologies Inc. An example is illustrated in Figure 14-11 along with current and voltage amplifiers that are used in testing power system relays.

Figure 14-11: A typical real-time digital simulator from RTDS Technologies Inc.

The other is the HYPERSIM version of a real-time digital simulator which is marketed by TransÉnergie Technologies Inc. It is available in two formats: SGI (UNIX based) and PC Cluster (LINUX based). The simulators utilize parallel processors techniques. Each format uses a central server node and other processors as computation nodes. The server node does both functions: computational as well as distribution to the other nodes.

14.5 PRESENT AND FUTURE TRENDS

With the increasing capabilities of digital computers and DSPs, the analog simulator has evolved gradually first to a hybrid and now to a totally digital implementation.

These trends are visible in replacing key components from analog to digital versions to minimize set up times e.g. micro-machines to replace physical machines [8]. In the 1970s, analog computers were used to solve the machine equations in real-time. These models are now totally replaced with a DSP version eliminating adjustments and setup times. Furthermore, different machine models can easily be pre-programmed and re-used many times.

Improvements have taken place also in the valve models used in the simulators. Previously, these were composed of a single thyristor to model a large thyristor valve. By using digitally controlled transistors, the behavior of the valve can be better represented in terms of recovery and protection characteristics. Furthermore, these valves can also represent the self-commutating switches like GTOs and IGBTs which are now required for newer VSCs.

As far as real-time digital simulator approach is concerned, continual developments in the following areas are to be expected:

- Lower cost,
- Reduced space requirements,
- Improved robustness (repeatability) and portability,
- Modularity, and
- User-friendliness.

One aspect that will continue to improve will be the calculation time step. With this, the detail of individual system models will also improve.

14.6 ACKNOWLEDGEMENT

This chapter is dedicated to my many colleagues at the IREQ Simulation Group. I thank also the Manitoba HVDC Research Centre and RTDS Technologies Inc. for allowing me the use of their documents and images to illustrate this chapter.

14.7 REFERENCES

[1]. Proceedings of ICDS'95, 5-7 April, Texas A & M, Texas, 1995.

[2]. Proceedings of ICDS'97 Conf., Montreal, 28-30 May 1997.

[3]. R.Kuffel, et al., "RTDS a fully digital power system simulator operating in real-time," ICDS'95, 5-7 April, Texas A & M, Texas, 1995.

[4]. P.Mercier et al., "Real-time digital simulation of power systems at Hydro-Quebec", ICDS'95, 5-7 April, Texas A & M, Texas, 1995.

[5]. J.C.Soumagne et al., "Development of the IREQ Simulator," ICDS'97 Conf., Montreal, 28-30 May 1997.

[6]. J.Belanger and L.Vaughan, "Hydro-Quebec Technology for the simulation of large and complex transmission systems", 8th CEPSI Conference, 5-9 Nov. 1990, Singapore.

[7]. C.Gagnon, V.K.Sood, J.Belanger, A.Vallee et al, "Hydro-Québec Power System Simulator", IEEE Canadian Review, No. 19, Spring-Summer 1994. pp 6-9

[8]. V.Que-Do and A.O.Barry, "A Real-Time Model of the synchronous machine based on Digital Signal processors", IEEE Trans. on Power Systems, Vol. 8-1, Feb. 1993, pp 60-66.

[9]. J.Le-Huy, J.C.Soumagne, Digital Real-Time simulation of transmission lines using parallel processors", IMACS-TCI Conference, Montreal, July 7-9, 1993.

[10]. C.Gagnon, P.Gravel, "Extensive evaluation of high performance protective relays for the Hydro-Quebec Series compensated network, "IEEE PES 94 Winter Power meeting 015-8, New York, Jan-Feb, 1994.

[11]. N.Mohan et al "Simulation of Power Electronic and Motion Control Systems - An Overview", Proc. of the IEEE, Vol. 82., No, 8, August 1994. pp 1287-1302

[12]. L.Dubé, H.W.Dommel, "Simulation of control systems in an electromagnetic transients program with TACS", Proc. IEEE PICA Conf., pp. 266-271, May 1977.

[13]. L.Dubé, I.Bonfanti, "MODELS: A new simulation tool in the EMTP", Euro-

pean Transactions on Electrical Power Engineering, Vol. 2, no. 1, pp. 45-50, Jan/Feb 1992.

[14]. L.Dubé, I.Bonfanti, M.T.Correia de Barros, V.Vanderstockt, "Using the simulation language MODELS with EMTP", Proc. 11th Power Systems Computation Conference, pp. 1063-1069, Aug 1993.

[15]. H.W.Dommel, "Digital computer solution of electromagnetic transients in single-phase and multiphase networks", IEEE Trans. Vol. PAS-88, April 1969, pp 388-399.

[16]. H.Dommel and W.S.Meyer, "Computation of electromagnetic transients," Proc. IEEE Vol. 62, July 1974, pp 983-993.

[17]. D.A.Woodford, A.M.Gole and R.W.Menzies, "Digital simulation of DC links and AC machines", IEEE Trans. Vol. PAS-102, No.6, June 1983, pp1616-1623.

Chapter 15

Modern HVDC - State Of The Art

15.1 INTRODUCTION

This chapter will look at modern trends in HVDC transmission in the present decade 2000-2010.

15.2 PAST DECADE VERSION

The driving forces for the developments of HVDC Converter terminals in the past decade were increased performance, increased reliability, reduced losses, higher overload capacity and better filtering with lower audible noise requirements. All of these requirements led to increased costs for HVDC transmission systems. The industry matured over this period and was characterized by the following features:

- **Valves:** Typical of the state-of-the-art valves during this period was the ±500 kV water-cooled valve for indoor utilization, having a 12 pulse, 3 quadri-valve configuration which was suspended from the ceiling for seismic reasons,
- **Converter Transformers:** These were three 1-phase winding transformers which were mounted close to the valve-hall with protruding bushes,
- **AC Filters:** These were mainly of the conventional, passive double-tuned and high-pass filters type with internal fused capacitors and air-cored reactors. Reactive power control was offered by switched capacitors,
- **DC Filters:** These were of the passive type with either air or oil cooled reactors. The DCCTs were of the zero-flux type, and

- **DC Controls:** These were mainly digital, but with some analog parts for the protection circuits and firing units. The controls were housed in the control room and were duplicated for hot standby.

15.3 PRESENT DECADE VERSION

The new trends in the present decade are being led by a commitment to reduce costs so that dc transmission can become competitive with ac transmission. These cost reductions are coming about due to reduced engineering costs as modular, standardized and re-usable designs are being employed. Furthermore, developments of the past decade in the areas of digital electronics and power switches have borne fruit.

15.3.1 Thyristor Valves

The ratings of thyristors has progressed almost linearly since the first thyristor valve became available in the early 1970s. Today, thyristor ratings of 8-9 kV and 1500 kW on silicon wafers of 150 mm diameter are commercially feasible (Figure 15-1). This has led to a dramatic decrease in the number of series connected thyristor elements comprising a valve, thus simplifying the design and reducing the power losses.

The thyristors can be either light or electrically triggered. It is claimed that Light-Triggered Thyristors (LTT) will offer performance and cost advantages in the future (Figure 15-2) by eliminating the high number of components in the electronic firing unit. Monitoring and protection features are also incorporated in these devices.

The valves are now of the air-insulated type and can be housed in outdoor units or modules with one valve per module. This will improve reliability and reduce costs due to the omission of large valve halls.

An important development in the usage of outdoor valves is a composite insulator which is used as a communications channel for the fibre optics, cooling water and ventilation air between the valve unit and ground (Figure 15-3).

An outdoor valve of this type has been in operation at the Konti-Skan I station since 1992 for 275 kV dc voltage (Figure 15-4).

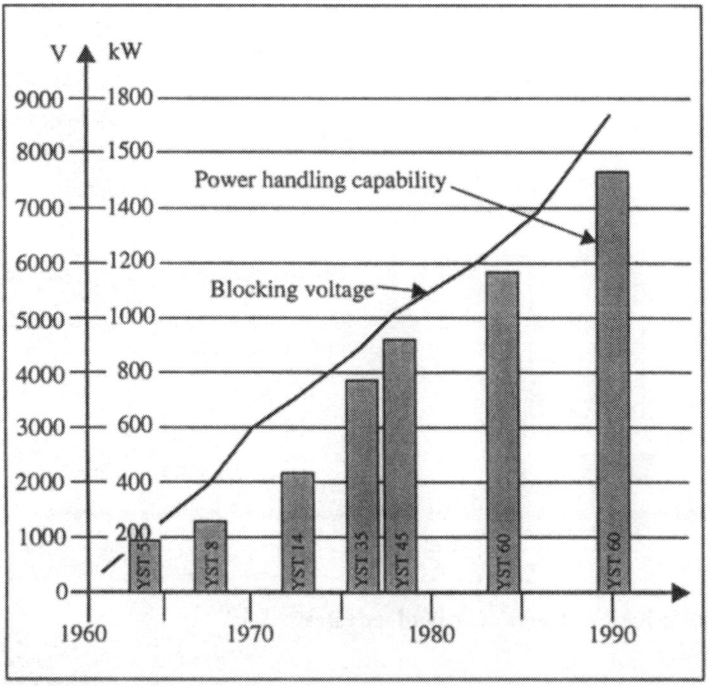

Figure 15-1: Development of blocking voltage and power handling capacity for HVDC thyristors

Silicon wafer and construction of the LTT. The light guides appear in the bottom right hand corner.

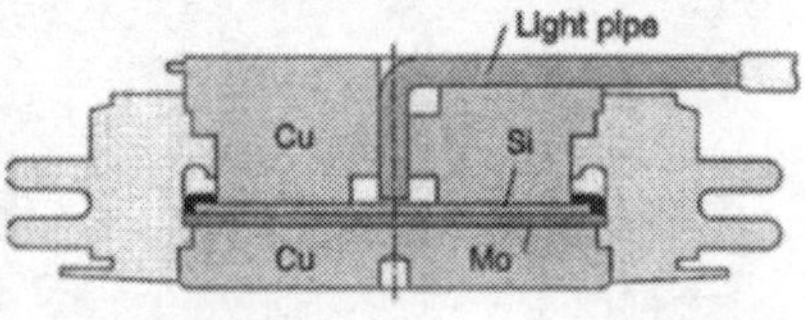

Cross section of the LTT with the light pipe entry.

Photos courtesy of Siemens.

Figure 15-2: Wafer and construction of the LTT thyristor

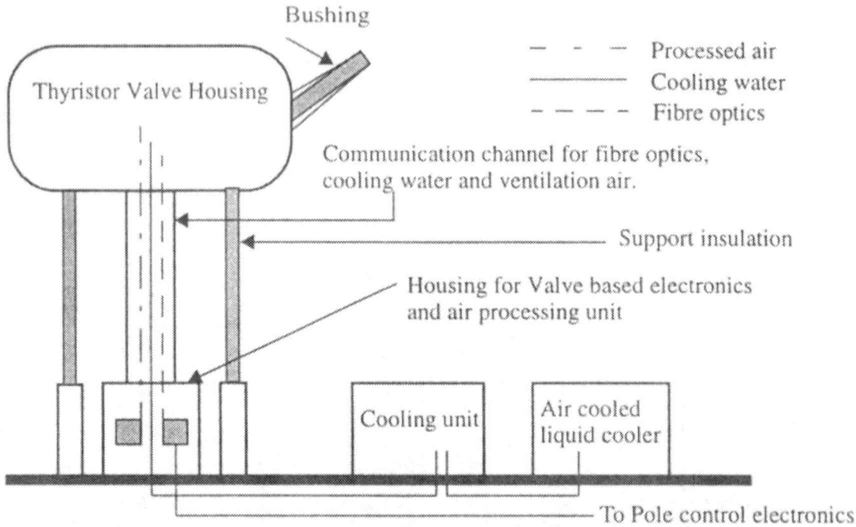

Figure 15-3: Basic elements of an outdoor valve

Figure 15-4: Outdoor valve in operation at Konti-Scan I station [ABB]

15.3.2 Self-commutated Valves

The increased interest in Voltage Source Converters (VSCs) has been a result of the development of self-commutated switches at increased power ratings. These switches now permit the use of sophisticated algorithms for deriving sinusoidal output waveforms from the VSCs for controlling active-reactive power and the generation-absorption of harmonics. A comparison of the power switches now being developed is made in Table 15-1.

Table 15-1: Comparison of power semi-conductor devices

	Thyristor	GTO *	IGBT *	SI *	MCT *	MOSFET *
Max. Voltage rating (V)	9000	6000	1700	2500	3000	1000
Max. Current rating (A)	4000	6000	800	800	400	100
Voltage blocking	symmetric/ asymmetric	symmetric/ asymmetric	asymmetric	asymmetric	symmetric/ asymmetric	asymmetric
Gating	pulse	current	voltage	current	voltage	voltage
Conduction drop (V)	1.2	2.5	3	4	1.2	resistive
Switching frequency (kHz)	1	5	20	20	20	100
Development target max. voltage rating (V)	10,000	10,000	3,500	5000	5,000	2000
Development target max. current rating (A)	8,000	8,000	2,000	2,000	2,000	200

* GTO - Gate Turn-Off Thyristor,
IGBT - Insulated Gate Bipolar Transistor,
SI - Static Induction Thyristor,
MCT - MOS Controlled Transistor, and
MOSFET - MOS Field Effect Transistor.

15.3.3 Active Filters

An active filter is a device that injects harmonic current at the same magnitude into the line in opposite phase so as to negate the influence of the harmonic. These devices become prominent in recent HVDC installations due to the following:

- Stringent requirements from the utilities for filtering harmonics,
- Availability of fast, economic power switches so that PWM VSC converters at high power and low losses are possible, and
- Increased capability of DSPs so that complex algorithms can be cheaply implemented.

It is feasible to have active filters on either the ac or dc sides of the HVDC converter. However, due to the combined requirements of reactive power supply and ac harmonic filtering on the ac side of the converter, the usage of active filters on the ac side has been somewhat muted as compared to the dc side application where reactive power supply is not required.

Figure 15-5: Active dc filter contained in a pre-fabricated module

Theoretically, either a current source or voltage source converter can be chosen for an active filter function, although VSC converters are more common.

15.3.3.1 AC Side Of The Converter

Conventional ac filters supply both reactive power to the converter and function as harmonic filters. With active filters the supply of reactive power, although feasible, is somewhat limited due to the cost factor associated with a large dc capacitor. The primary function of the active filter is therefore to counteract the influence of the harmonics, and let the reactive power be handled by other techniques.

15.3.3.2 DC Side Of The Converter

On the dc side of the converter, the active filter is concerned with mitigating the harmonics from entering the dc line and generating telephonic interference.

Figure 15-6 shows the cost of the active dc filter compared to the passive filter as a function of the equivalent disturbing current. The cost of the passive filter increases dramatically as the disturbing current is reduced. However, the cost of the active filter remains virtually constant since it consists of a simple passive filter in series with an active part which covers the frequency range.

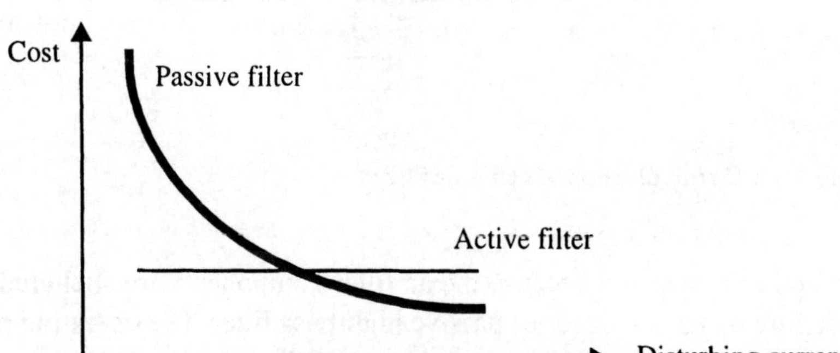

Figure 15-6: Cost of dc filter with disturbing current

As with all active filters, the possibility of either series or parallel (shunt) operation is possible. The choice between them is made by consideration of the following:

- Measuring points,
- Harmonic source,
- Integration of the active filter within the system, and
- Regulation aspects from the circuit point of view.

Due to the cost of components required to operate in series with a high voltage dc line, the shunt option (shown in Figure 15-7) has been developed.

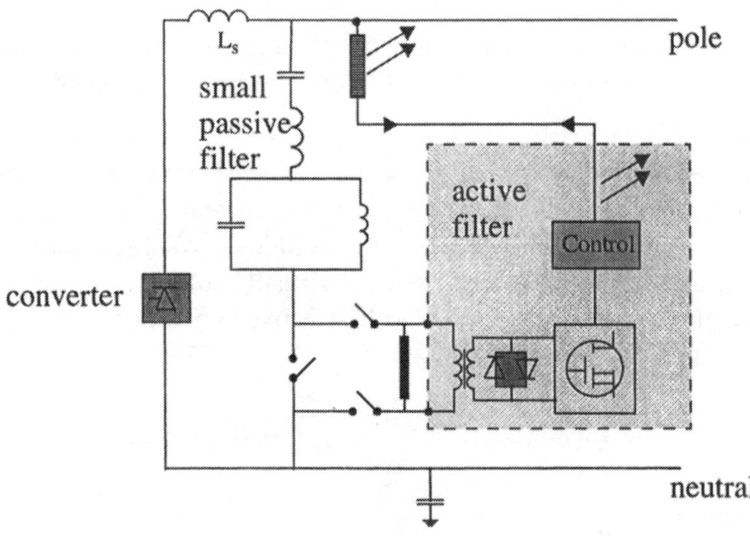

Figure 15-7: Circuit diagram of active dc filter

This option is selected because the ac filter components are sheltered from the dc line by the small series passive high-pass filter. The operating principle of an active dc filter is shown in Figure 15-8. A prototype of this type of filter was installed in the Konti Skan II terminal in 1991. The dc filter injects a current in the filter in opposition to the harmonic current from the HVDC converter so as to cancel it. The active filter is connected at the low voltage end of a small passive dc filter connected between the pole and neutral buses.

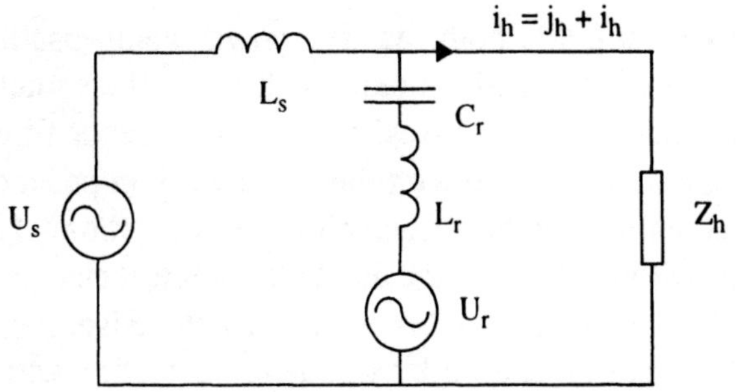

Figure 15-8: Simplified circuit diagram for HVDC Converter with active dc filter

When the active part was connected, a significant reduction in the harmonic current in the dc line was achieved (Figure 15-9).

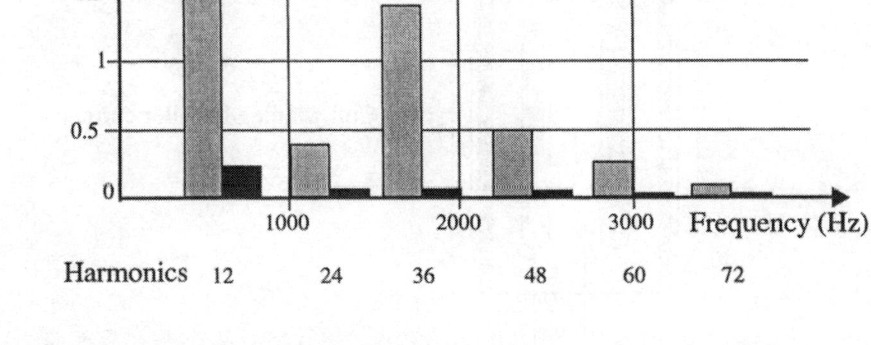

Figure 15-9: Harmonic current on the dc line with passive and active dc filters

15.3.4 Tunable AC Filters

Traditional ac filters provide both reactive power compensation and harmonic filtering. Using high quality factors with these filters maintains a low filter impedance which reduces losses, and provides better filtering. However, due to a possible resonance with the ac network impedance, it is also necessary to maintain a certain bandwidth of operation. The network impedance varies within a wide range due to topological changes in the network caused by lines connections/disconnections etc. Also, the detuning of conventional filters can be caused by frequency deviation and component variation e.g. capacitance changes due to varying temperature. These variations can result in inadequate filtering and reduce the performance of the filter.

Using continuously tunable filter inductors is not a new idea, and was attempted many years ago. The idea was abandoned due to operational and maintenance difficulties in maintaining tuned filters caused by temperature fluctuations and aging of equipment.

Various techniques for achieving tuning are possible i.e. using either mechanical or electronic taps on reactors etc. However, the active tuning of a reactor by means of the orthogonal magnetization of an iron core inside the reactor provides an attractive solution with no moving parts. The magnetization of the core can be controlled by a direct current (Figure 15-10).

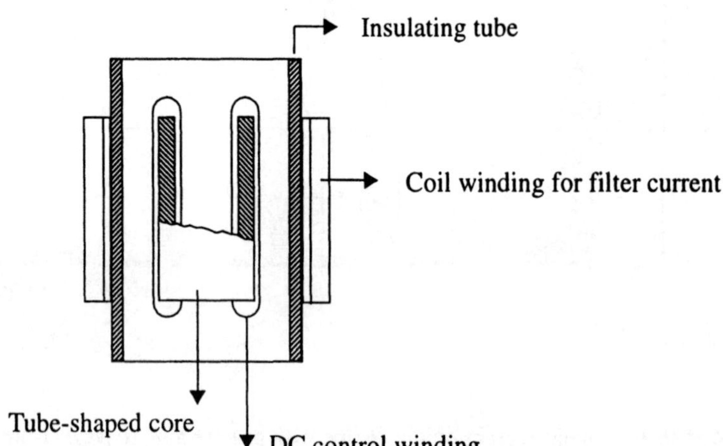

Figure 15-10: Orthogonal magnetizing of ConTune Filter reactor

15.3.5 AC-DC Measurements

An optical current transducer is now used to replace earlier more complicated measuring equipment. The new transducer is based on a high-precision shunt in series with the line (Figure 15-11).

Optical Current Transducer (OCT)

Block Diagram, OCT

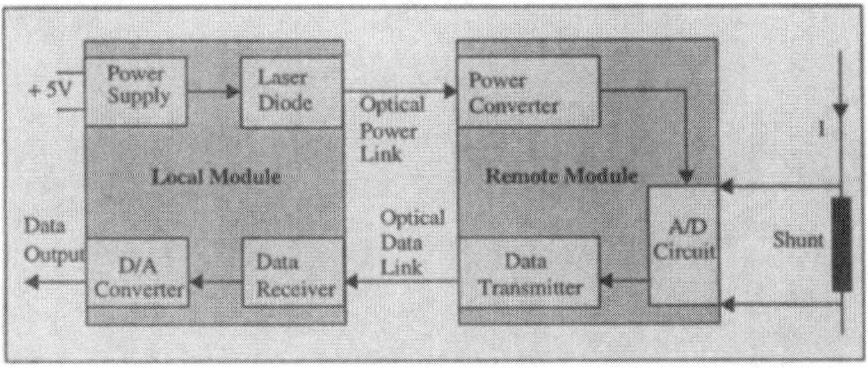

Figure 15-11: Optical Current Transducer (OCT) [ABB]

The A/D converter at high potential is powered by a light (laser) source sent to the OCT via a fibre optic link from the control room. The measured current is transmitted in digital form to the detection equipment in the control room. The length of the fibre optic link can be upto 300 m long. The accuracy of the measurement technique is better than 0.5% in the frequency range from dc to 7 kHz. The use of an optical link provides better reliability, compact design and lower costs.

15.3.6 Digital Signal Processor (DSP) Controllers

DSPs have grown exponentially in the last decade and now can have a calculation capacity of 2000 million instructions per second (MIPS). This has led to improved reliability and fully hot standby redundant control systems. Even the final stronghold of analog controls i.e. main converter firing control systems are now fully digital. The number of cubicles required in the control room have decreased leading to lower costs for buildings. Additionally, improved control algorithms, lower maintenance, fault diagnosis features and online monitoring are now being offered by manufacturers.

15.3.7 Compact Station Design

The station layouts of a typical 2000MW HVDC station designed and built in the 1980s and a modern layout are illustrated in Figures 15-12 and 15-13 (figures are to scale). The conventional station has an area of 300 x 300 m and the modern station has an area of 100 x 200 m i.e. a 25% reduction.

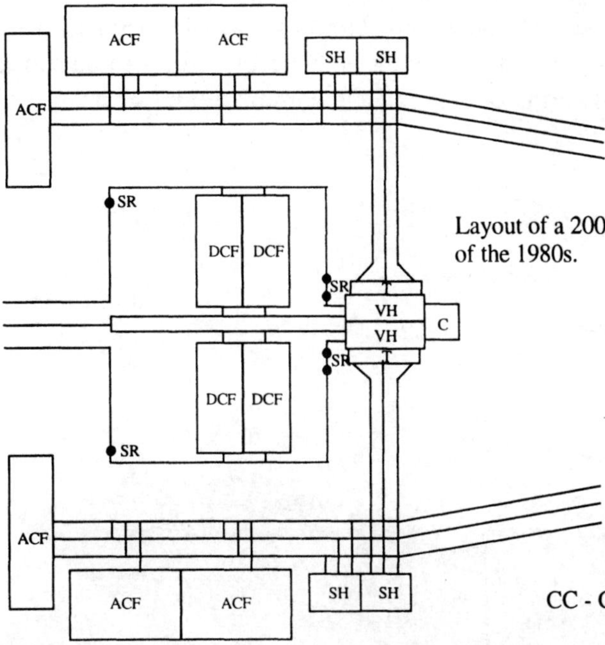

Layout of a 2000 MW bipolar station of the 1980s.

ACF - AC Filter
DCF - DC Filter
VH - Valve Hall
VY - Valve Yard
SH - Shunt Capacitor
SR - Smoothing Reactor
C - Control Building
CC - Control & Auxi. Modules
T - Transformers

Figure 15-12: Comparison of a 2000 MW HVDC station of the 1980s with a modern design [1,3]

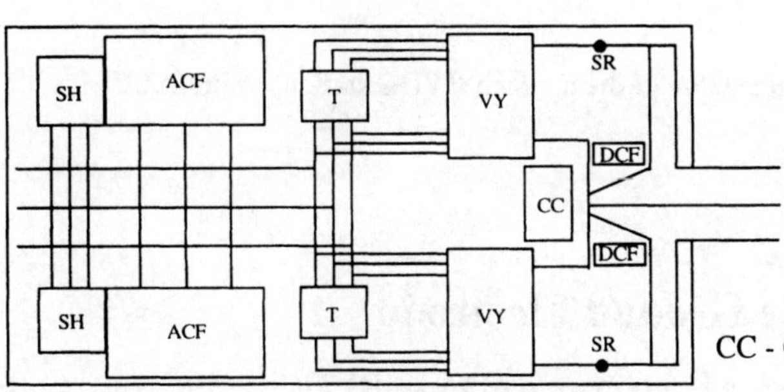

ACF - AC Filter
DCF - DC Filter
VH - Valve Hall
VY - Valve Yard
SH - Shunt Capacitor
SR - Smoothing Reactor
C - Control Building
CC - Control & Auxi. Modules
T - Transformers

Proposed Layout of a modern 2000 MW bipolar station.

Figure 15-13: Comparison of a 2000 MW HVDC station of the 1980s with a modern design [1,3]

An artist's impression of the modern station is shown in Figure 15-14. The most important factors in the size reduction of future HVDC terminals are active ac-dc filters, outdoor valves and the housing of control and auxiliary systems in pre-fabricated modules. No buildings will be needed, and all controls will be factory tested. Site work is, therefore, likely to be substantially reduced, with lowering of erection and commissioning costs.

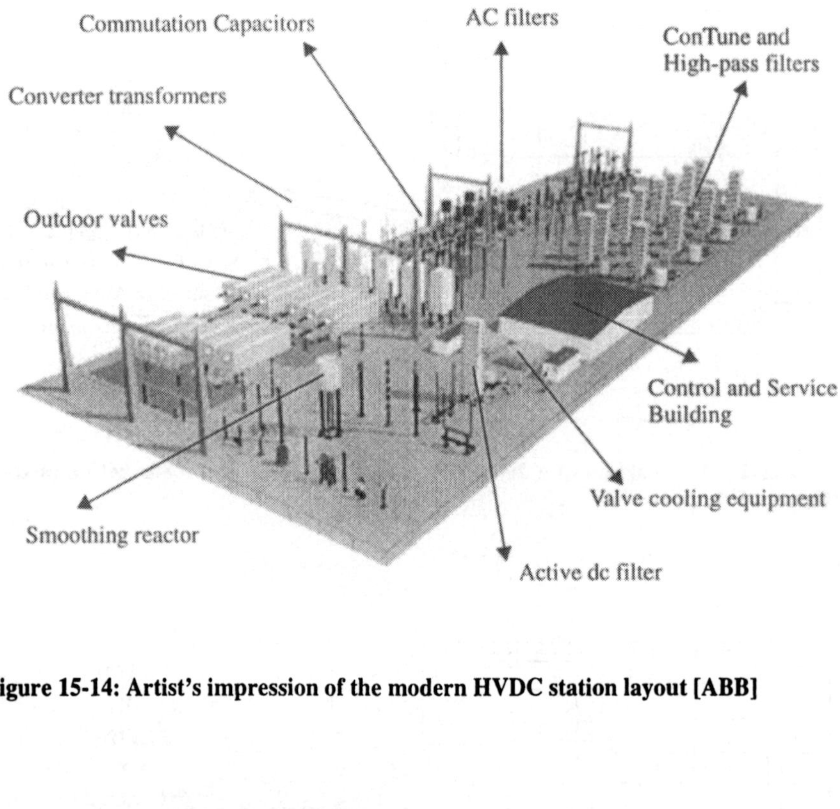

Figure 15-14: Artist's impression of the modern HVDC station layout [ABB]

15.3.8 Deep Hole Ground Electrode

The ground electrode is an important part of an HVDC system, and contributes to the overall cost of the terminal. In a submarine cable link or a terminal close to the sea, it is feasible to build compact shore or sea electrodes. However, often a terminal will be in a location where access to a good grounding point is not conveniently available. In such cases, a land electrode requires a large area and must be located where the earth resistivity is low. Under such conditions, a deep hole ground electrode (Figure 15-15) can be placed in a earth layer with low resistivity resulting in low electric potentials and potential gradients at the surface.

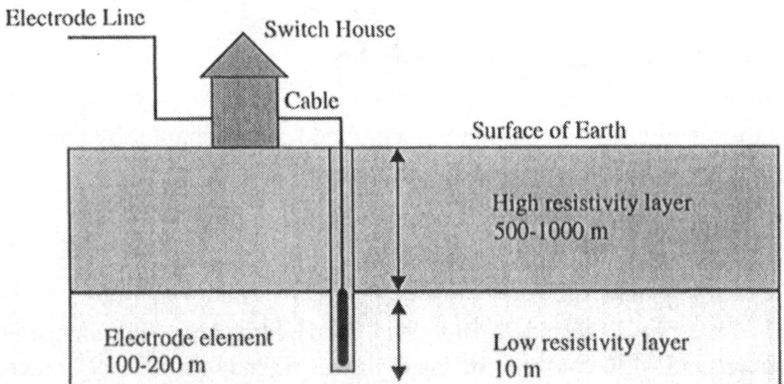

Figure 15-15: Deep earth electrode [7]

The advantages of such an electrode are:

- Allows the electrode to be closer to the converter station,
- Usage of a shorter line with reduced power loss,
- Reduced interference and reduced risk of lightning strikes,
- Easier to find a suitable electrode site, and
- Enhanced possibilities to operate the dc link in mono-polar mode.

15.4 CONCLUDING REMARKS

The future use of HVDC systems will also consider forced commutation techniques which offer a greater control of active and reactive power, with the reduced generation of harmonics. The use of self-commutated switches will eliminate the problem of commutation failures. With the increased rating of power switches, the requirement of expensive transformers may also be affected.

Since HVDC terminals are now considered part of the FACTS domain, an awareness of the deregulated utility market will lead to an integrated approach for meeting new challenges. The use of wind, photo-voltaic, micro-turbines and fuel cells to provide new sources of power will provide new opportunities for HVDC applications.

15.5 ACKNOWLEDGEMENTS

The author is grateful to information supplied for this chapter by many individuals at the manufacturers ABB and Siemens.

15.6 REFERENCES

[1]. L. Carlsson, G.Asplund, H. Bjorklund, and H. Stromberg, "Recent and future trends in HVDC converter station design", IEE 2nd International Conference on Advances in Power System Control, Operation and Management, December 1993, Hong Kong.

[2]. W. Zhang, G.Asplund, A.Aberg, U.Jonsson and O.Loof, "Active dc filter for HVDC system: A test installation in the Konti Skan dc link at Lindome Converter Station", IEEE Transactions on Power Delivery, Volume: 8, Issue: 3, July 1993, pp 1599 - 1606.

[3]. Carlsson, L.; Asplund, G.; Bjorklund, H.; Stomberg, H., "Recent and future trends in HVDC converter station design", Advances in Power System Control, Operation and Management, 1993, 2nd International Conference on APSCOM-93, 7-10 Dec 1993, vol.1, pp 221 - 226.

[4]. T.Jonsson and P.Bjorklund, "Capacitor commutated converters for HVDC", ABB documents.

[5]. T.Holmgren, P.Hidmann, U.Jonsson. S.Valdemarsson and O.Loof, "A test installation of a self-tuned ac filter in the Konti Skan 2 HVDC Link", ABB documents

[6]. G.Asplund, I.Blidberg and O.Loof, "Outdoor thyristor valve for HVDC", ABB documents.

[7]. D.Karlsson, S.Nyberg, J.Sjodin and L.Eriksson, "Deep hole ground electrode for HVDC transmission", ABB documents.

Index

About the Author

Vijay Kumar Sood obtained his B.Sc.(1st Class Hons.) from University College, Nairobi in 1967 and his M.Sc. degree from Strathclyde University, Glasgow in 1969. He obtained a Ph.D. degree from the University of Bradford, England in 1977.

From 1969-76, he was employed at the BR Railway Technical Centre, Derby, UK. In 1976, he emigrated to Canada and was employed as a Researcher at IREQ (Hydro-Québec) in Montreal. In 1984, he also joined Concordia University, Montreal as an Adjunct Professor.

He is a Member of the *Ordre des ingènieurs du Québec*, a Senior Member of the Institute of Electrical and Electronic Engineers (IEEE), a member of IEE (UK) and a Fellow of the Engineering Institute of Canada. He has received numerous awards; among them are the 1998 Outstanding Service Award from IEEE Canada, the 1999 *Meritas Award* from the *Ordre des ingènieurs du Québec*, the IEEE Third Millennium Award from IEEE in 2000, the IEEE Regional Activities Board Achievement Award for 2001, and the Engineering Institute of Canada's Canadian Pacific Railway Engineering Medal for 2002. He is presently the Managing Editor of the *IEEE Canadian Review* (a quarterly journal of IEEE Canada); his term ends in 2005. He serves as an associate editor of the journal Control Engineering Practice for Pergamon Press, a Director and Treasurer of IEEE Montreal Conferences Inc. and a Director of the IEEE Canadian Foundation (till 2006).

Dr. Sood has worked on the modeling of electrical power systems and their controllers for over 28 years. His research interests are in the monitoring, control and protection of power systems using artificial intelligence techniques. He has published over 70 articles and written two book chapters. He has supervised 14 post-graduate students and examined 17 Ph.D. candidates from universities all over the world.

Lightning Source UK Ltd.
Milton Keynes UK
14 July 2010

156962UK00007B/166/P

9 781402 078903